INSIGHTS FROM INSECTS

"[T]he most thoughtfully balanced presentation of insects and our battles against these remarkable creatures available to the general reader. Waldbauer brilliantly presents insect pests as sources of both genuine inspiration and human ingenuity."

Jeffrey A. Lockwood, author of *Locust: The Devastating Rise and Mysterious Disappearance of the Insect that Shaped the American Frontier*

"[F]ascinating. . . . Gil Waldbauer is a natural storyteller who has pulled together facts from a multitude of sources. . . . Both informative and enjoyable for the naturalist, layperson, student of evolution, and nonprofessional and professional entomologist."

Robert J. Marquis, Professor, Department of Biology,
University of Missouri-St. Louis

"Dr. Waldbauer's book is both interesting and well written. It gives the reader the opportunity to learn positive facts about so-called pest bugs. I highly recommend it."

Richard "Bugman" Fagerlund, syndicated columnist and author

"Insights indeed! In this good-humored and instructive little book university scientist Gilbert Waldbauer's primary focus is on what he terms 'bad bugs' and the usual suspects. . . ."

Gerry Rising, State University of New York
Distinguished Teaching Professor Emeritus,
Buffalo News "Nature Watch" Columnist

"[P]acked with information, yet written in an easy style that keeps the reader turning the pages. Each example . . . provides the reader with an interesting insight into how the natural world operates. . . . This is one of few biology books written so accessibly that readers find they have learned a lot without realizing it."

Dr. Whitney Cranshaw, Extension Specialist of Entomology,
Department of Bioagricultural Sciences and Pest Management,
Colorado State University

"An informative book on how a few pest insects have taught us much about insect behavior and natural selection."

William H. Luckmann, Professor Emeritus,
Illinois Natural History Survey, University of Illinois

"An entomological 'Most Wanted List' that contains fascinating lessons from the insect world."

James K. Wangberg, PhD, author of *Do Bees Sneeze? and Other Questions Kids Ask about Insects* and *Six-Legged Sex: The Erotic Lives of Bugs*

INSIGHTS FROM INSECTS

What Bad Bugs Can Teach Us

Gilbert Waldbauer

59 John Glenn Drive
Amherst, New York 14228-2197

Drawings by Meredith Waterstaat; photographs by Philip Nixon and James Sternburg

Published 2005 by Prometheus Books

Inquiries should be addressed to
Prometheus Books
59 John Glenn Drive
Amherst, New York 14228–2197
VOICE: 716–691–0133, ext. 207
FAX: 716–564–2711
WWW.PROMETHEUSBOOKS.COM

09 08 07 06 05 5 4 3 2 1

Library of Congress Cataloging-in-Publication Data

Waldbauer, Gilbert.
Insights from insects : what bad bugs can teach us / Gilbert Waldbauer.
p. cm.
Includes bibliographical references and index.
ISBN 1–59102–277–0 (pbk. : alk. paper)
1. Insect pests. 2. Insects. 3. Insect pests—Ecology. 4. Insects—Ecology. I. Title.

SB931.W249 2005
632'.7—dc22

2004026928

Printed in the United States of America on acid-free paper

To the late Dottie Nadarski

Sorely missed by all of us in the Department of Entomology

CONTENTS

	Acknowledgments	11
	Introduction	13
1.	The Most Dangerous Insects *Mosquitoes*	17
2.	Evolution in Action *House Fly*	37
3.	What Darwin Wished He Knew Drosophila	49
4.	Natural Selection Outflanks Farmers *Corn Rootworms*	61
5.	How a Species Becomes Two Species *Fruit Flies*	69

6. Guaranteeing Descendants: The Numbers Game
Aphids 81

7. Guaranteeing Descendants: The Role of Parental Care
Tsetse Fly 91

8. Surviving Winter as a Sleeping Egg
Evergreen Bagworm 99

9. Escaping Predators by Deception
Black Swallowtail Butterfly 113

10. Why Insects Are Such Picky Eaters
Cabbage White Butterfly 129

11. "Nutritional Wisdom"
Corn Earworm 143

12. Invaders from Abroad
Gypsy Moth 155

13. An American Saves the French Wine Industry
Grape Phylloxera 171

14. An Insecticide "Creates" New Pests
Codling Moth 183

15. From Low- to High-Tech Controls
European Corn Borer 195

16. The Demise of DDT
Japanese Beetle 207

17. Their Passing from the Agricultural Scene
Chinch Bug 219

18. Synchrony with the Seasons
Hessian Fly 233

19. An Insect to Control Another Insect
Cottony Cushion Scale 243

20. Extermination by Subverting the Sex Act
Screwworm Fly 253

Epilogue 265

Notes 273

Index 293

ACKNOWLEDGMENTS

I owe a great debt of gratitude to the many friends and colleagues who helped me by giving generously of their time, advice, encouragement, and knowledge: James Appleby, May Berenbaum, Stewart Berlocher, Bryan Flood, Rosanna Giordano, Fred Gottheil, Jeffrey Grannet, Christina Grozinger, Larry Hanks, Karen Hogenboom, Elizabeth Jeffery, Eli Levine, William Luckmann, Robert Novak, Rose Reynolds, Hugh Robertson, Gene Robinson, James Sternburg, Paul Strode, Rick Weinzierl, and Arthur Zangerl.

Special thanks to Phyllis Cooper, who read and improved most of the chapters, and to David Nanney, who, with his unexcelled understanding of Mendelian genetics and its history, contributed significantly to the third chapter. My book benefited from the painstaking editing of Mary Read and Linda Regan. The drawings are the products of Meredith Waterstaat's wonderful talent. The photographs are the work of ace photographers Philip Nixon and James Sternburg.

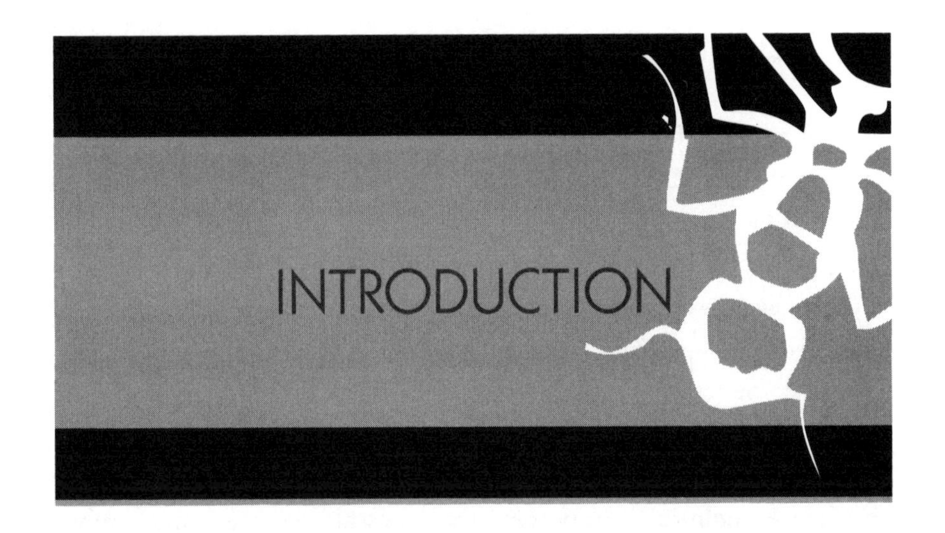

INTRODUCTION

At dusk a large, narrow-winged hawk moth hovers like a hummingbird in front of petunias in your garden, pollinating them as she uncoils her long tongue and probes for nectar. She and others like her, fueled by the sugar in the nectar, will spend the night laying eggs one at a time on plants favored by her species: plants of the nightshade family, among them wild plants such as purple nightshade itself and cultivated plants such as tobacco and tomato. Several of these moths lay eggs on tomato plants in your garden. At first, the caterpillars that hatch from the eggs, known as tobacco hornworms, are so small that their feeding goes unnoticed, but during the last week of their three- to four-week caterpillar stage, they will strip whole leaves from the plants, consuming nearly 90 percent of their total food intake and growing to be almost four inches long. In this situation the tobacco hornworm is indisputably a pest: it will destroy your tomatoes. Furthermore, all the subjects of the following insect stories—which I think of as verbal portraits of the insects and their way of life—are pests.

But why did I choose pest insects as the subjects of these portraits, as exemplars in these explorations of the amazing lives and ways of insects? One reason, of course, is that, because they directly affect our lives, most people are more or less familiar with one or more of them and may even want to know more about them. But the main reason is that, generally speaking, science has learned much more about these few irksome species than about the overwhelming majority of innocuous insects. Entomologists have taken to heart the familiar aphorism "know your enemy" as it pertains to pest insects. Because they are economically important there has always been much more money available to support research on pest insects than insects that are not pests. Consequently, not only the applied research but also even much of the basic research done by entomologists has focused on insects that annoy us or cause economic losses. Furthermore, because pest insects are readily available and relatively well known scientifically, they often serve as "laboratory animals" for basic research on genetics, behavior, and physiology.

What is it that makes an insect a pest? In his 1989 textbook *Entomology and Pest Management*, Larry Pedigo offers a concise definition: "Pest species are those that interfere with human activities."[1] (But keep in mind that less than 2 percent of the 900,000 known insects can reasonably be considered to be pests.) In 1915 Stephen Alfred Forbes, a great pioneer entomologist and founder of the University of Illinois's Department of Entomology, wrote a more philosophical statement:

> The struggle between man and insects began long before the dawn of civilization, has continued without cessation to the present time, and will continue, no doubt, as long as the human race endures. It is due to the fact that both men and certain insect species constantly want the same things at the same time. Its intensity is owing to the vital importance to both of the things they struggle for. We commonly think of ourselves as lords and conquerors of nature, but insects had thoroughly mastered the world and taken full possession of it long before man began the attempt. . . . They

> have disputed every step of our invasion of their original domain so persistently and so successfully that we can even yet scarcely flatter ourselves that we have gained any very important advantage over them.[2]

An old saw known to gardeners has it that a weed is a plant growing where we do not want it. And so it is with pest insects. Whether or not an insect is being a pest depends upon the place, the season, and the circumstances; an insect is not always bad. For example, a tiny caterpillar that feeds only on plants of the thistle family does farmers a service when it attacks harmful weeds such as the alien bull thistle. But this insect, the aptly named artichoke plume moth, is a pest when it infests cultivated artichokes, which are just huge, unopened thistle buds. Termites are important members of the forest environment, feeding on dead wood, recycling it by returning its elements to the soil. But a colony of them is certainly not welcome if, doing what comes naturally, they set up housekeeping in a wooden building, particularly if it is your home.

The pest insects included here are but a small sample of the great diversity of the many insects with which we share our planet, but they are revealing examples of the many different ways in which all insects cope with the problems of surviving, wresting a living from their environment, and bearing offspring to perpetuate their genes. Nevertheless, every one of them is unique and interesting in its own right, as are all organisms. But, each of them is also an especially apt, perspicuous, and interesting example of the lifestyle of an insect; of a fundamental aspect of biology such as evolution, genetics, ecology, or food procurement; or of the many ways in which people are threatened by insects and what can be done to contend with them.

As you read on you will find out what the house fly tells us about natural selection and evolution, and what we can learn about ecology from the chinch bug, Hessian fly, and codling moth. You will read the story of how the humble little fruit fly, *Drosophila*, became the "laboratory animal" of choice for genetic research that yielded and still yields findings that apply to all organisms, including humans. The portrait of the black swallowtail butterfly, the parent of the parsley caterpillar, tells us how it, although perfectly edible, tricks insect-eating birds into sparing it.

These insect portraits provide the surprising answers to some equally surprising questions. Why do so many insects starve to death rather than feed on a plant not on their menu? How did Henry Ford help to control house flies? Why do some flies have a uterus and a mammary gland? Why do insects with little or no protection not freeze solid when the temperature drops far below the freezing point? How did an American entomologist save the vineyards of France from inevitable destruction? This book will answer many other questions, and, I sincerely hope, give you a better and more appreciative understanding of insects. They are, after all, members of the same creation that produced us. It also behooves us to know that, without them, most of Earth's ecosystems would collapse, and life as we know it could not continue.

1

THE MOST DANGEROUS INSECTS

Mosquitoes

In 1960 William Downes, a fellow entomologist, and I were collecting insects in Texas. One afternoon we camped on a deserted Gulf beach on the Bolivar Peninsula, only a few miles northeast of Galveston. We made a fire of driftwood, heated our meal of canned food, and watched the sun go down over the marsh on the landward side of the peninsula. As it began to darken, we saw what looked like a large cloud of smoke rising from the marsh. We soon realized that it was an immense swarm of countless millions of salt marsh mosquitoes departing their larval habitat in search of blood meals. The offshore breeze was blowing them toward us. They soon arrived and proceeded to plague us almost beyond bearing. Hundreds hovered in front of our eyes, and a few bit us despite the repellent we had smeared on ourselves. When we took off our boots before taking refuge in our screened tent, squadrons of them instantly landed on our repellent-free ankles and bit us mercilessly.

Mosquitoes are a nuisance. Worse, mosquitoes can be a threat to life. Certain mosquitoes—by no means all of them—have a greater pernicious impact on people than does any other insect or group of insects. Keep in mind that most of the 3,000 known species, over 150 in North America, do not affect people. Some of the others are indeed the most dangerous of our insect enemies. We all know how annoying mosquitoes can be. Along the coast of New Jersey, for example, the biting salt marsh mosquito—"the New Jersey state bird"—can make life miserable. Inland, a variety of other mosquitoes often take the joy out of a backyard barbecue, especially after heavy rains have flooded their breeding places, which are the bottom lands of rivers and ephemeral woodland pools. The scientific names conferred on some species reflect the torment they can cause: *Psorophora horrida*, *Culex perfidiosus*, *Mansonia perturbans*, *Aedes vexans*, and *Aedes tormentor*.

The annoyance caused by mosquitoes, although sometimes severe, is nothing compared to the widespread suffering and millions of deaths caused by the diseases transmitted by a few of them. There are many mosquito-borne diseases of humans, but considering just a few of the more troublesome will make the point: malaria, caused by an amoeba-like protozoan, although now rare in North America, is arguably the most devastating of the infectious diseases worldwide; West Nile fever, caused by a virus, has been spreading rapidly throughout the United States since its first appearance in the Western Hemisphere in New York in 1999; yellow fever, another viral disease, still present in the tropics, was in the nineteenth century the scourge of port cities from New Orleans to Philadelphia, and for decades put on hold the digging of the Panama Canal.

Knowing them and their ways is, as with all pest insects, a prerequisite for rationally coping with mosquitoes. Understanding them in order to better control them has been the primary goal of many entomologists, epidemiologists, and physicians. In *Mosquito: A Natural History of Our Most Persistent and Deadly Foe*, Andrew Spielman and Michael D'Antonio refer to mosquitoes as the "magnificent enemy."[1] They are indeed magnificent creatures and would be worthy of our interest even if they were not among our deadliest foes. They are true flies that have been finely honed by the evolutionary process to thrive or at least survive almost anywhere on Earth, not only in temperate climates and tropical jungles but even in deserts, the frozen tundra of the Arctic, at the edge of the sea, and as high as 14,000 feet in the mountains.

A fundamental reason for the success of mosquitoes is that like most other insects, among them beetles, wasps, bees, butterflies, and moths, mosquitoes undergo a *complete metamorphosis* that includes four stages: egg; larva (the growing stage); pupa (the transition stage), in which wingless larvae metamorphose and become adults (the final winged, reproductive stage). This transition has made it possible for larvae and adults to follow different evolutionary paths. The larvae are specialized for feeding and growing, the adults for mating and distributing their eggs. Larval and adult mosquitoes and larval and adult butterflies are as different as snakes and birds. Larval butterflies—caterpillars—have been referred to as a digestive system on a caterpillar tread and adults as flying machines devoted to sex.

About 15 percent of the known insects, among them grasshoppers, cockroaches, dragonflies, and cicadas, undergo a *gradual metamorphosis* with only three life stages: egg; the growing stage (called the *nymph* to differentiate it from a larva); and the adult or reproductive stage. Nymphs and adults, with the exception of some aquatic insects, look very much alike except for their size and the state of development of their wings. They are also generally similar in behavior. Nymphal and adult grasshoppers, for example, live in the same habitat, are both active jumpers, and eat the leaves of plants. Except for the fully developed wings and genitalia of the adult, neither life stage has become explicitly specialized for either growth or reproduction.

All mosquitoes are aquatic in the larval and pupal stages and aerial in the adult stage. The legless larvae, commonly known as *wrigglers*, live in still or sometimes slowly flowing water, usually where there is

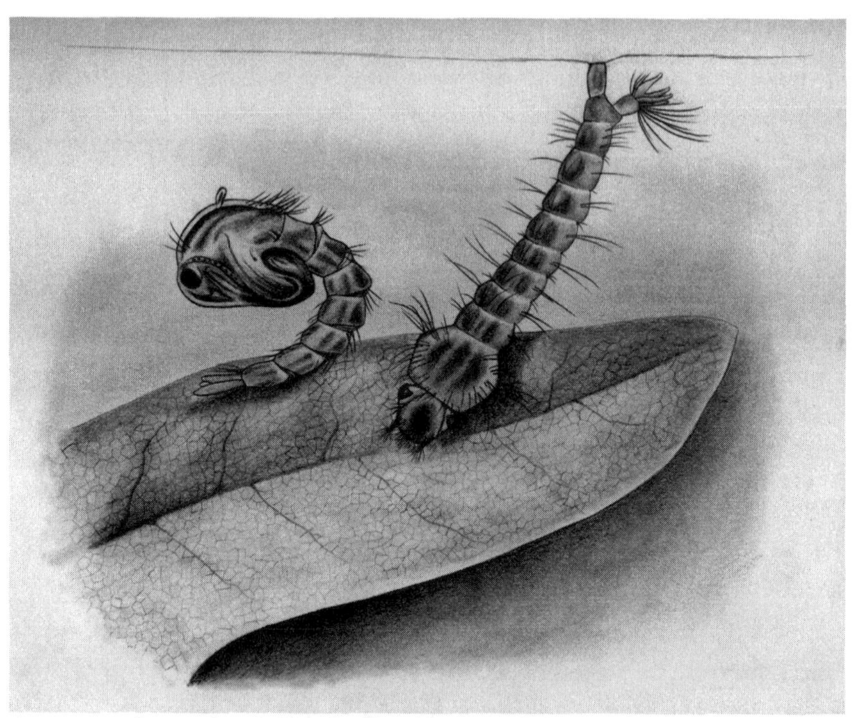

A mosquito larva taking a breath through its snorkel in a woodland pool. A pupa is to its left.

abundant aquatic vegetation in which they can hide from fish and other enemies. Some live in tree holes—water-filled rot cavities in the trunks or branches of trees—or even in small containers such as vases in a cemetery or in clogged rain gutters of houses; others find a home in the water cupped by the leaves of air plants that cling to the trunk or branches of tropical trees; and a few live in the water in the pitfall traps of insect-eating pitcher plants. Because they have no gills for extracting dissolved oxygen from the water, most mosquito larvae must come to the surface to get oxygen directly from the air. The majority of them have at their tail end a long air tube—a snorkel—which penetrates the surface film of the water as they hang head down from it. Instead of an air tube, larvae of the genus *Anopheles*, the malaria mosquitoes, have just a flat plate bearing a pair of spiracles (porelike openings to the respiratory system) on the undersurface of the tip of the abdomen. Stretched out horizontally, the *Anopheles* larvae hug the surface film as they press their spiracles up against it. Species of the genus *Mansonia* obtain oxygen in a unique way. The air tube has become a sharp piercing organ, which is stabbed into the air chambers in the underwater portion of a marsh plant such as the cattail.

The larvae of most mosquitoes feed on plankton—microscopic organisms such as minute, one-celled green algae and copepods, tiny relatives of the crabs—that they filter from a current of water created by and directed to their mouth opening by the rhythmical fanning of a pair of hairy brushes, one on either side of the mouth. Some larvae feed as they hang from or lie under the surface film, and others feed near the bottom or even take their microscopic prey from the surface of the bottom muck. A few larvae have biting mouthparts and are predators that eat mosquito larvae, usually of other species, but they will also turn on their own kind.

About seven days after they hatch from the egg, mosquito larvae, which have grown to full size after molting their skins three times, undergo the molt to the pupal stage. Unlike the pupae of almost all other insects, mosquito pupae, commonly known as *tumblers*, are mobile and very active. Although they have no legs or mouthparts for feeding, they are energetic swimmers that can, at the threat of danger, speedily race from the water surface to the bottom and then back up again. They are comma-shaped and very different in appearance from either larvae or adults. The head and thorax are united, forming a large bulbous mass, the head of the comma. The upper surface of the thorax bears a pair of short, trumpet-shaped breathing tubes that can penetrate the surface film of the water. The abdomen, the tail of the comma, is slender and flexible, and has at its tip a pair of finlike appendages that act as paddles when the abdomen is repeatedly and rapidly flexed to scull the pupa through the water with a tumbling motion. Like the larvae, mosquito pupae are sensitive to disturbances in the water and have eyes that perceive shadows overhead. Make a shadow by waving your hand above a group of larvae and pupae and they, always on the alert, will instantly dash down to the safety of the bottom. This fright response might well save them from the piercing beak of predaceous water bugs or the gnashing mandibles (chewing organs) of predaceous water beetles that swoop down from the air to dive into the water. Generally speaking, the pupal stage lasts for only two to three days. At the end of that time, the pupa comes to the top and lies just under the surface film. Its skin splits and the adult within steps out onto the surface, rests for a half hour as its skin hardens, and then flies away.

Adults of most of the 3,000 or more known species of mosquitoes do not feed on blood, but rather subsist only on nectar or other plant juices. The several species of one small genus, *Harpagomyia*, obtain food in a very unusual way; they trick ants into regurgitating and then sip up the regurgitant, which is likely to contain nectar or honeydew (the sugary excrement of aphids). In a letter to the Entomological Society of London, C. O. Farquharson of Ibidan, Nigeria, describes how both females and males waylay ants:

> You know how worker ants stop each other and exchange a little regurgitated food, a momentary transaction almost, both passing quickly on their way. The mosquitoes do exactly the same. They will drop downwards just over an ant that is hastening along in the usual way. The ant may stop and give alms to the beggar, passing on a moment or two later as if it had just met a friend, and the mosquito flies up and down again till another obliging ant is met. At times the selected ant simply ignores the mendicant, but shows no resentment, nor does the mosquito press his or her attentions.[2]

Females of the relatively few bloodsucking mosquitoes attack humans and other mammals, birds, reptiles, frogs, or even insects. The males make do with nectar and other plant exudates. The females consume these substances too, mainly as a source of energy, but some require the protein in blood to produce eggs. Mosquitoes that do not feed on blood carry over sufficient protein from the larval stage.

Females "bite"—take blood meals—mainly at dawn, dusk, during the night, and sometimes on cloudy days. A few species feed in broad daylight. When a female takes a blood meal, a troughlike protective scabbard, a greatly modified and lengthened lower lip, bends back out of the way so that the other mouthparts, exquisitely adapted for piercing and sucking, can penetrate the skin of the victim. The piercing "needle" is actually a long tube composed of six long, separate, stiletto-like stylets wet with saliva that adhere to each other, although two of them can slide forward and backward on the others. The bite usually causes little or no pain because the "needle" is more like a hypodermic needle than a sewing needle or a pin. Rather than being painfully jabbed into the skin like a pin, which is actually relatively blunt and tears the flesh, the mosquito's mouthparts, sharper than the razorlike tip of a hypodermic needle, gently and painlessly cut through the skin as the sawlike and exceedingly sharp tips of the two sliding stylets gently slice into the skin sliding in and out. The mosquito's piercing tube is traversed by two channels, a wide one though which the blood is sucked into the digestive system and a narrow one through which saliva containing an anticoagulant can be injected into the vertebrate host.

The males of many species of mosquitoes gather together in large mating swarms, usually at dusk. In 1906 Frederick Knab vividly described mosquitoes swarming in Urbana, Illinois,[3] and for the first time reported that swarming brings the sexes together. A swarm of thousands of mosquitoes formed over his head when he walked into a recently cut field of corn adjacent to a polluted stream squirming with wrigglers. Another swarm hovered over a nearby corn-stook (a teepee-like shock of cut cornstalks), and "a round of the field showed that each corn-stook had its swarm of mosquitoes, and furthermore, single stalks that remained standing had small swarms dancing over them. . . ." Knab also notes that "always the mosquitoes gathered over some prominent object such as a tree or a projecting branch, a bush, a cornstook, or a person. In this last case the swarm would move with the person and the only way to get rid of it was by passing under some taller object where the swarm would then remain." In every case the swarms consisted almost exclusively of males, which all faced into the breeze as they hovered and danced over their chosen landmark. When he swept his net through a swarm, he caught 897 males but only 4 females. Females are attracted to the swarm, enter it, and are immediately pounced on by a male. Each pair leaves the swarm to copulate nearby and after that the female flies off and the male returns to the swarm. Males recognize females by the distinctive tone of their rapidly beating wings. Indeed, caged males are attracted to a vibrating tuning fork if the pitch is right. In a clever experiment, caged males clustered on a cloth that hid a vibrating tuning fork and tried to copulate with each other and even with the cloth.

After being inseminated, the females lay eggs. The way in which they are deposited differs with the species of mosquito. Females of the genus *Culex* and some other genera form rafts of hundreds of eggs; the rafts are about the size of a caraway seed and float on the surface of the water. Others, among them the many species of *Anopheles*, lay their eggs singly on the water surface. The eggs are kept afloat by bubble-like air cells on the shell. *Aedes* females and those of some other genera lay their eggs in dry places that may eventually become covered with water. For example, *Aedes aegypti*, the yellow-fever mosquito, lays its eggs just above the water surface in rot holes in trees or in artificial containers, and *Aedes vexans*, one of the most abundant of the pest mosquitoes in North America, lays them in the moist soil of river bottoms, ephemeral pools, or other places that are periodically flooded.

The eggs of *Aedes vexans* and other "flood water" mosquitoes can survive in the soil for two years or more, and will hatch after they are covered with water. According to William Horsfall,[4] the eggs hatch only after the following sequence of events: completion of the development of the embryo, a period of drying before flooding, being covered by water at a favorable temperature, and finally a sharp drop in the amount of dissolved oxygen in the water. The latter requirement is particularly interesting because it prevents the larvae from emerging from the egg before there is sufficient food in the water to sustain them. Because the planktonic organisms that the larvae eat use up dissolved oxygen, a sharp decrease in its concentration signals the presence of abundant food for the larvae. In the laboratory, *vexans*' eggs will not hatch in distilled water, which, of course, contains no plankton. However, if the water is seeded with plankton, the eggs hatch after the planktonic organisms have multiplied. They will also hatch in uncontaminated distilled water if the concentration of dissolved oxygen is artificially lowered by bubbling nitrogen gas through the water.

How do such delicate creatures as mosquitoes survive the winter? Many of them, notably the floodwater species, spend the winter as diapausing eggs. (Diapause, a winter "sleep," is the insects' version of hibernation.) Others, among them the species of the genera *Anopheles* and *Culex*, overwinter as fertilized females hiding in such protected sites as hollow trees, caves, or in the case of *Culex pipiens*—the house mosquito—in human habitations, especially cellars, or in dry places in storm drains. A few, including the pitcher plant mosquito and some that breed in flooded cavities in trees, survive the winter as larvae frozen in the water in their breeding sites.

Today we know that mosquitoes and other bloodsucking insects are transmitters—*vectors*—of disease-causing viruses, bacteria, protozoans, and nematode worms. But not much more than a hundred years ago this crucially important fact was not known. Malaria (it means "bad air" in Italian) was thought to be caused by miasmas emanating from swamps and marshes. Physicians believed that yellow fever was caused by filth or caught from handling contaminated bedding and clothing. Some few speculated that bloodsucking insects transmit disease, but there was no proof of this until 1878 when Patrick Manson, an English parasitologist, reported that the nematode worms which cause a disease of humans called filariasis are transmitted by a mosquito.[5] Similar discoveries followed, and by 1900 it was well established that insects can transmit diseases, including malaria and yellow fever.

In the eighteenth and nineteenth centuries, epidemics of yellow fever periodically struck port cities along the Atlantic and Gulf coasts from Philadelphia to New Orleans. A city might be free of this disease for years, and then one summer yellow fever would suddenly appear. In those days, yellow fever was a terrifying mystery. Almost nothing was known about it—only that it caused the skin to turn yellow, that its symptoms (high fever, jaundice, vomiting, and hemorrhages) were agonizing, and that it was often fatal. But since then much has been learned. For one, we know that when there are no human cases, the yellow fever virus persists in troops of monkeys that live high above the ground in the leafy canopy of tall trees in the Amazonian forest. Mosquitoes of the genus *Haemogogus*, which breed in water-filled tree holes, hollows in the trunk or branches in the canopy, transmit the virus from monkey to monkey but not to other animals. As epidemiologists put it, this is the "reservoir" of the yellow fever virus, the place where it smolders when it is not blazing in the human population.

How did the virus get from the Amazon jungle to faraway New Orleans or some other American port city? William Horsfall, one of my mentors in graduate school, regaled his classes in medical entomology with a hypothetical but plausible scenario: it all began when a woodcutter felled a tall tree in the jungle, bringing its lofty canopy with many of its residents, including *Haemogogus* mosquitoes, down to ground level. As he lopped branches from the fallen tree, he was bitten by a canopy-dwelling mosquito that inoculated him with the yellow fever virus. Not long after, he guided a raft of logs down the Amazon

River to the port of Belem, where he intended to sell them. By the time he arrived in Belem, the six-day incubation period of the disease had passed and he fell sick. As he lay suffering, he was bitten by another kind of mosquito, *Aedes aegypti*, that sucked up the virus with his blood and spread it to residents of Belem. (In most areas where it occurs, *aegypti*, known as the yellow-fever mosquito, is associated with people, breeding in water-filled artificial containers and taking its blood meals from people.) Soon there was a yellow fever epidemic in Belem. Then an American sailor who came ashore was inoculated with the virus when he was bitten by a female *aegypti*. He showed no symptoms until six days after his ship had sailed. If he had died or become immune to the virus after recovering, that would have ended the story—except for an important circumstance. There was a flourishing colony of yellow-fever mosquitoes aboard the ship, breeding in water barrels kept on deck to slake the crew's thirst. The sailors did not mind a few wrigglers in the water. (I have heard that they welcomed them as a sign that the water was pure.) The mosquitoes on board spread the virus from the ailing sailor to his shipmates. By the time the ship reached New Orleans, the members of the crew were either dead or sick with yellow fever. New Orleans had its own population of the yellow-fever mosquito, which is generally common in warm, humid areas throughout the world—especially in seaports. It was not long before yellow fever spread from the sailors to the people of the city and the epidemic was underway.

West Nile fever, caused by a mosquito-borne virus related to the yellow fever virus, made its first appearance in the Western Hemisphere in the borough of Queens in New York City in 1999. First isolated and identified in Uganda in 1937, the West Nile virus was subsequently found to occur in much of Asia and Africa, with occasional incursions into Europe. But how did it get to New York? Spielman and D'Antonio noted that it could have been introduced in only three ways: in a sick person, an infected mosquito, or an infected bird.[6] A human source is improbable because human blood generally contains too little virus to contaminate a mosquito. It may have come to New York in mosquitoes that stowed away on an airplane. But the most likely scenario is that it arrived in illegally imported birds that had not been quarantined before entering the country. There is even a small chance that it was brought in by a bird that strayed on its own to our shores from Europe. I know from my own birding experiences that a few strays from Europe show up in North America virtually every year.

In 1999, in New York City and some nearby areas, hundreds of birds, mainly crows, starlings, and house sparrows but also others such as blue jays, hawks, and owls, were killed by the West Nile virus, as were some exotic birds in the Bronx Zoo and several horses on Long Island. There were 62 human cases and seven deaths. Uriel Kitron, a pathologist in the University of Illinois College of Veterinary Medicine, told me that only three years later, the virus had spread from Nova Scotia to California and that human cases were most numerous in Illinois and Louisiana. As told to me by Robert Novak, a medical entomologist at the Illinois Natural History Survey, in North America, the West Nile virus is now known to infect 157 species of birds and 36 species of mosquitoes. It has also been found in some mammals other than humans, among them horses, dogs, wolves, bats, squirrels, chipmunks, and others. The virus is a serious threat to wild birds. Many people, including experienced birders, have noted that birds, particularly crows, have recently been much less numerous than they usually are. Their observations are supported by the 100,000 or more infected dead birds, from tiny chickadees to large crows and great horned owls, that have already been found.[7] Rare birds—species with small populations, such as whooping cranes, peregrine falcons, and piping plovers—could even be threatened with extinction. We can only hope that North American birds will soon develop some immunity, as seems to be the case with European birds. According to Novak,[8] in 2002 Illinois led the nation in the number of human cases and had the record number of 64 fatalities. About 1,000 horses were found to be infected, and about half of them died.

The West Nile virus is apparently spread by migratory birds that cross and recross the country as they move north in spring, south in autumn, and then back again. A bird that gets infected with the virus just before leaving an area may still be alive and infectious when it later stops to rest and feed in an area that the West Nile virus has not yet reached. Over 200 species of birds migrate between their breeding ranges in Canada and the United States and their winter ranges in the Caribbean and Central and South America. Many herons stray northward in autumn and then fly back south. In autumn Bohemian waxwings, lovely fawn-colored birds, searching for trees with berries move from far western North America to as far east as New York and then return. A few other birds on their way south from their usual breeding range in eastern North America make a wrong turn and end up as far west as California, where they delight the local birders but might also be carrying the West Nile virus.

Which mosquitoes transmit the virus from birds to humans, and where does the virus lurk in winter when neither birds nor humans become sick? It is generally agreed that the virus is spread to people mainly by the common house mosquito, *Culex pipiens*, in the north and by *Culex quinquefasciatus* in the south. These two species suck blood from both birds and people, although the former prefers birds and the latter prefers humans. The virus has been found in other mosquitoes, including the floodwater species *Aedes vexans*, by far the most numerous of the pest mosquitoes in many parts of North America. No one knows if it transmits the virus from birds to people. The virus survives the winter in mosquitoes. As Spielman and D'Antonio said, in 1999 the big question was whether or not the West Nile virus would survive the winter in New York.[9] The answer was found when it was discovered in the bodies of several adult female *Culex pipiens* (house mosquitoes) overwintering in the shelter of an underground tunnel.

Malaria, debilitating and often fatal, is arguably the worst of the infectious diseases of humans. Four types of human malaria are caused by four protozoans of the genus *Plasmodium*, single-celled relatives of the amoeba, that can survive only in humans or the *Anopheles* mosquitoes that transmit them from person to person. Humans are not the only animals that suffer from malaria. At least a hundred other species of *Plasmodium*, also transmitted by mosquitoes, often not *Anopheles*, cause malaria in monkeys, rodents, birds, and reptiles. Malaria often kills these animals. For example, bird malaria drove to extinction or near extinction many dozens of native Hawaiian birds, birds that evolved on the isolated Hawaiian Islands and are found nowhere else in the world. Birds that migrate from North America to spend the winter in Hawaii had always been infected with bird malaria, but it did not spread to the native Hawaiian birds because there were no mosquitoes in Hawaii—not until 1826, when the crew of a whaling ship emptied their water barrels, alive with mosquito larvae, into a stream on the island of Maui.

The various species of *Plasmodium* have complex life histories that can be completed only if they parasitize both a mosquito and a vertebrate host. The human/*Anopheles* life cycle is well understood and will serve as an example. A life stage of the *Plasmodium* known as the *sporozoite* is injected into a person by a mosquito, enters a liver cell, and reproduces asexually by division. Its numerous offspring, *merozoites*, enter red blood cells and produce more merozoites, which are released when the red blood cells burst. They go on to produce several generations of this life stage. (The simultaneous release of hordes of merozoites into the blood causes the chills and fever that are characteristic symptoms of human malaria.) Some merozoites produce male and female forms that remain in red blood cells, which may be ingested by a mosquito. In the mosquito's stomach, these male and female forms unite, the *Plasmodium*'s version of sex, and after several transformations produce numerous sporozoites that wander in the blood, concentrate in the salivary glands, and begin another cycle if the mosquito, in taking a blood meal, injects them into a human with its saliva.

Today human malaria occurs in tropical areas of Central and South America, Africa, Asia, and the East Indies, but until the mid-twentieth century it was far more widespread, established in virtually all subtropical and tropical areas and some temperate areas. In Europe it occurred throughout Italy, the Balkans, and along the Mediterranean, Atlantic, and Baltic coasts from Italy and sometimes to St. Petersburg in Russia. There were also temporary outbreaks as far north as 64° north latitude in Sweden, only about 160 miles south of the Arctic Circle. In the United States there was no human malaria in precolonial times. A 1947 publication of the Tennessee Valley Authority stated:

> Malaria was introduced into the United States in colonial days by emigrants from . . . the Old World. The importation of Negro slaves for a period of 200 years constantly replenished the supply of malaria carriers. [White slave traders probably also carried malaria.] Water was used for transportation and for power to a greater extent than prevails today, and settlements, as a matter of course, developed along streams. Colonial housing offered little or no impediment to the entrance of mosquitoes, and contemporary documents indicate that the inhabitants were constantly plagued by mosquito bites. The result was that malaria moved westward with the settlement of the country.[10]

By the end of the nineteenth century, malaria was present in isolated areas of the far west and most of the eastern two-thirds of the country from the Canadian border south to the Gulf of Mexico. Then the disease began to recede southward and by 1934 occurred mainly in the southeastern states from eastern Texas to Virginia, but was most severe along the Gulf and Atlantic coasts. Apparently malaria receded because breeding sites for the mosquitoes were drained for agricultural and industrial purposes, and because, with increasing prosperity, people lived in better housing less open to mosquitoes. By the middle of the twentieth century, malaria was all but obliterated from most of the developed world, but was still a common and deadly presence in third world countries.

Malaria has taken a great toll in human lives and caused a great deal of misery. In 1955 there were 200 million cases and 2 million deaths worldwide. When Illinois was first being settled in the mid-nineteenth century, before the marshy prairies had been drained, malaria was rife and the state was known as the graveyard of the nation. Between 1870 and 1874, according to Horsfall's medical entomology text,[11] a US Army unit of 171 men stationed in Louisiana reported 1,229 cases of malaria—more cases than men because the men were infected and reinfected repeatedly. The effects of malaria on local economies, wrote Spielman and D'Antonio, were severe.[12] In North Carolina, for example, the efficiency of workers in textile mills was cut in half during the four months of the malaria season.

In 1955 the World Health Organization (WHO) began a worldwide campaign to eradicate malaria by greatly minimizing contact between humans and the *Anopheles* mosquitoes that transmit it. People are most likely to be bitten by anophelines as they sleep during the night in homes that are not protected by screens, which is often the case in developing countries. As part of the WHO project, the mosquitoes, which rest on walls and ceilings during the day, were killed by spraying these surfaces with DDT, which left a toxic residue that persisted for as long as six months and that was absorbed through the mosquitoes' feet.

For a time, the eradication campaign was a great success.[13] Malaria was eliminated from 36 countries, and its rate of occurrence was reduced in many others. The eradication campaign failed and was eventually ended in 1969, but WHO still supports some efforts to reduce mosquito populations. Malaria had been all but eliminated from Sri

Lanka before the 1970s. Today over a million cases are reported from this island nation each year. There have been massive resurgences of malaria in India, Pakistan, Africa, and elsewhere. In 1995 WHO reported that worldwide there are now 300 to 500 million cases per year and from 1.5 to 2.7 million deaths per year, the great majority of them in Africa. According to one estimate, in Africa about 250 children die of malaria every hour. Malaria is again at least as prevalent as it was before 1955. It kills twice as many people as AIDS.

Why did the eradication campaign fail? The basic reason, as Robert Metcalf explains, is that *Anopheles* mosquitoes became resistant to DDT and the other inexpensive insecticides that were first used to control them.[14] Different insecticides were substituted, but the mosquitoes soon became resistant to them and to many others as one failed insecticide after another had to be replaced. The campaign became more and more costly because each substitute insecticide was usually more expensive than the one it replaced. Substituting malathion for DDT increased costs by about fivefold, and then switching to the carbamate insecticide propoxur increased costs by about twentyfold. As Metcalf points out, this combination of resistance and economics was responsible for the return of epidemic malaria.[15] To make matters even worse, the malaria-causing plasmodia have, in some areas, become resistant to the synthetic compound chloroquine and other antimalarial drugs.

Malaria is steadily increasing among tourists and other travelers who visit malarial areas of the tropics. Anyone who plans to visit such areas should consult a physician about taking preventive doses of antimalarial drugs and should also prepare to protect himself or herself against mosquitoes. Some returning tourists and visitors or immigrants from malarial areas bring with them to the United States plasmodia in their blood. Since *Anopheles* mosquitoes that are competent vectors of plasmodia are almost everywhere in the United States, there has been, for the first time in decades, the local transmission of malaria in this country.

People repeatedly infected with malaria develop a partial immunity to the disease. They can become infected and have plasmodia in their blood but develop only mild symptoms or no symptoms at all. A wry twist of nature causes natural selection to favor and tend to perpetuate a deadly hereditary disease of humans, sickle-cell anemia, because it confers a degree of immunity to the deadly *falciparum* form of malaria. Sickle-cell anemia is expressed in humans who inherit the sickle-cell gene from both parents. These individuals are sick most of their lives, generally die before the age of twenty, and seldom bear children. Consequently, you would think that natural selection would soon eliminate the sickle-cell trait, but this has not been so because individuals who inherit the sickle-cell gene from only one parent usually do not suffer the symptoms of this disease, are much less likely to die from malaria than people who do not have the sickle-cell trait, and, therefore, can pass the sickle-cell trait to their offspring.

A watercolor painting of a mosquito—much enlarged—hangs in my office. I look at it off and on as I write. The mosquito stands gracefully on long, slender legs, and her thorax is ornamented with stripes of iridescent, sapphire-blue scales, giving credence to her name, *Uranotaenia sapphirina*. She is among the most beautiful of the mosquitoes, but all mosquitoes are, if not beautiful, at least delicate and graceful, belying the inescapable fact that the few, not including *Uranotaenia*, that transmit diseases are humanity's deadliest insect foes.

These few transmit not only yellow fever, West Nile fever, and malaria, but a great many other diseases of humans and other vertebrates. Among the many is the eastern encephalitis virus, which causes a rare but often fatal disease in humans, and more often kills horses and certain birds. Encephalitis is an inflammation of the brain that can kill or cause mental impairment. It is transmitted from birds to people by mosquitoes of the genus *Culiseta*. A disease mainly of infants and young children, it often results in mental retardation. Myxomatosis, a deadly viral disease of rabbits often transmitted by mosquitoes, was introduced into Australia to control an almost nationwide catastrophic plague of rabbits, descendants of European rabbits imported to improve hunting. So virulent is this disease that within three years it had spread throughout the infested areas of the continent and had virtually wiped out the rabbits. Caused by tiny filiarial Nematode worms (roundworms), human filariasis is transmitted from person to person by mosquitoes of the genus *Culex*. The worms cause lymph ducts to become blocked and swell, causing a symptom known as elephantiasis, a gross enlargement of arms, legs, female breasts and genitalia, and scrotums. I have seen a picture of a man whose scrotum is so hugely enlarged that he can move about only by resting it in a wheelbarrow.

Some uninformed people fear that the AIDS virus can be transmitted by mosquitoes. That could happen only if a mosquito, instead of sucking a victim's blood, injected the victim with blood from a person infected with AIDS—an unlikely and probably impossible occurrence.

What I have just said about mosquitoes raises some momentous questions that get at the very heart of our understanding of life on Earth. Why are there 3,000 different species of mosquitoes? For that matter, why are there 1,200,000 known species of animals and as many as 300,000 species of plants? How did mosquito larvae become such well-adapted aquatic creatures? How did some female mosquitoes become so adept at taking blood meals from animals, usually without getting slapped or whisked away? Then there is the question of how the complex three-way relationship between humans, mosquitoes, and *Plasmodia* came to be. The answer to all of these questions is that evolution, driven by natural selection, is constantly molding plants and animals to cope with the demands of their environment. In this way, new species are constantly arising, while others fall by the wayside and become extinct. Our exploration of the house fly, which comes next, shows us that this insect is not only an interesting animal in itself, but also provides us with an easily understood example of natural selection. In plain view and in less than two years, natural selection shaped house flies to survive a new and deadly threat that appeared in their environment.

2 EVOLUTION IN ACTION

House Fly

In the opening decade of the twentieth century, one of the first forest rangers in Montana ordered dessert in a local eatery near the Deer Lodge National Forest, pointing to the counter and saying, "I'll have a piece of that blueberry pie." The waitress fanned her hand over the pie to chase off a multitude of house flies and said, "That ain't blueberry; that's apple." The ranger was Aretas Saunders, who more than 30 years later taught me biology at Central High School in Bridgeport, Connecticut. I do not doubt his story. Books and articles from the early twentieth century report that house flies were a veritable plague, entering homes and businesses by the thousands, their numbers only slightly diminished by flyswatters and long strips of sticky "flypaper" that hung from the ceiling and soon became black with the bodies of trapped flies.

The house fly (*Musca domestica*) is one of those animals that, like several species of cockroaches, the bed bug, three species of clothes moths, the house mouse, and the Norway rat, is a pest associated with people all over the world and is at least partially dependent upon them. (Entomologists write the name of a real fly [order Diptera] as two words; as one word, for example, butterfly, if it's not a real fly.) This fly, wrote the British medical entomologist George Graham-Smith, "is probably the most widely distributed insect to be found; the animal most commonly associated with man, whom it appears to have followed over the entire earth. It extends from the subpolar regions to the tropics, where it occurs in enormous numbers."[1]

This fly is usually no more than a nuisance but may also transmit disease-causing viruses and bacteria, especially in underdeveloped countries where sanitation is poor. House flies may carry bacteria both externally and internally. When a fly walks on rotting garbage or feces, its feet and the rest of the outer surface of its body become contaminated with bacteria, as many as several million of them. Consequently, the fly will taint any food it walks on. Sucking moisture from garbage or feces, including those of people, contaminates the fly's digestive system with bacteria. If it next feeds on solid but soluble food, perhaps sugar in a bowl, it will dissolve it with saliva and contaminated regurgitant and then suck it up, but leave bacteria in the bowl.

Typhoid fever is probably the worst of the diseases that can be transmitted by house flies. In 1898, during the Spanish-American War, about a fifth of the American soldiers living in encampments came down with typhoid, and, as reported in Luther West's comprehensive book on the house fly,[2] this disease was responsible for 80 percent of all deaths among the troops. Many of the house flies that swarmed in mess tents and freely crawled on food bore traces of the white, powdered, quicklime that was frequently sprinkled on the excrement in pit latrines. The flies could have picked up the quicklime only by contacting human feces in latrines, feces likely to be contaminated with the bacterium that causes typhoid fever, which causes not only fever, but also intestinal inflammation, and severe diarrhea.

When horses were the usual mode of transportation, house flies were far more abundant than they are now—distressingly so—and it was estimated that about 90 percent of them developed in horse manure. It has been said, and not wrongly so, that Henry Ford did more to reduce house fly populations than has any other person. Ford's ingenious methods of mass production made it possible to manufacture automobiles so inexpensively that most people could afford to buy one. By the beginning of the 1930s, the Model T and the Model A Fords had replaced most horses, and no longer was there a pile of horse manure in almost everyone's backyard. When I was a small child in the 1930s in Bridgeport, Connecticut, I never saw anyone riding in a horse and carriage, but a few horse-drawn wagons were still used in the city by the iceman, the milkman, peddlers of vegetables, and the junk man who collected rags and wastepaper.

Despite Henry Ford, flyswatters, flypaper, and efforts to clean up dung piles and other breeding sites, house flies, although greatly diminished in numbers, were and still are abundant enough to cause problems, especially on dairy farms where manure accumulates and milk might be contaminated by the flies. Then DDT came on the scene. As Thomas Dunlap wrote in his comprehensive book on DDT, this white powder was thought to be "the atomic bomb of insecticides, the killer of killers, the harbinger of a new age of insect control."[3] A chlorinated hydrocarbon compound, the first of the synthetic organic insecticides, it had been synthesized by the German chemist Othmar Zeidler in 1874, but its insecticidal properties were first discovered in the 1930s by Paul Müller of the Swiss J. R. Geigy and Company. By 1942, during World War II, the US armed forces were using DDT to kill the scourge of malaria-transmitting mosquitoes on the South Pacific islands.[4] It became generally available for civilian use in 1945, and weak dilutions of it in water were sprayed to control many different pest insects, including the house fly. It was, for example, sprayed on the inner walls of dairy barns, where it persisted for as long as six months. Flies that landed on the walls, as most eventually did, absorbed tiny doses of DDT through their feet and soon died. It seemed like a miracle. After a few days it looked as though all the house flies were gone.

The miracle was short-lived. By 1946, only a year after DDT was first used in agriculture in the United States, house flies and other insects had become resistant to it and continued to become more resistant to it from year to year. Anthony Brown and R. Pal, insect toxicologists, cite data that show how resistant wild house flies collected on Illinois farms had become.[5] In 1945 a dose of as little as 0.18 microgram of DDT per gram of body weight—a really minuscule amount—was sufficient to kill the most susceptible half of a group of flies, but by 1951 as much as 125 micrograms per gram of body weight was required to achieve the same result. The dosage that would kill house flies had increased by a factor of almost 700 times in only six years! James Sternburg told me that the resistance of house flies to DDT can increase even more than that. One of the house fly colonies in his laboratory was kept in a cage whose walls, ceiling, and floor were always covered with a residue of DDT. These flies could walk on pure, undiluted DDT and survive; they were 1,000 or more times as resistant as flies from a susceptible colony that was scrupulously guarded against any contact with DDT. House flies are not unique in their development of resistance. Worldwide by 1990, about 500 different pest insects had become resistant to DDT and/or one or more of the many other synthetic insecticides that later became available.

James Sternburg, Clyde Kearns, and Herbert Moorfield, of the Department of Entomology of the University of Illinois, discovered that resistant flies, but not susceptible ones, contain an enzyme that is presumably there for some other purpose but which by happenstance is capable of detoxifying DDT by degrading it to DDE, which, though harmless to insects, is still toxic to birds and other vertebrates.[6]

How did house flies—and other insects, too—become resistant to DDT so quickly? The answer is that, although it seemed as though

DDT had killed all the flies in dairy barns and other places that had been sprayed in 1945, there were actually just a few surviving flies—so few that they went unnoticed. They survived because they were resistant to DDT. This mere handful of survivors became the parents of the next generation, and passed on to their offspring the genetically determined ability to detoxify DDT. In his book *Nature Wars*, Mark Winston writes that the development of resistance to an insecticide "is the same phenomenon as evolution by natural selection" except that insecticides, produced by humans, can be viewed as an artificial selective force.[7] DDT and other insecticides are especially rigorous selective agents, because they kill most of the target insects, sparing only the few resistant individuals. As Winston points out, the survivors multiply rapidly, and soon replace the susceptible individuals, because they have the great advantage of "growing and reproducing in an environment in which competition for food and other resources has been vastly diminished by the thinning power of pesticides."

Evolution is the central and unifying concept of biology: the science of life, the science through which we seek to understand ourselves and our fellow creatures, to know where we came from, what we are, and how we are inextricably bound to all other life on Earth. The driving force of evolution is natural selection, sometimes popularly known as the survival of the fittest. The success, or fitness of an animal, not usually achieved just by proficiency with fang and claw, is measured by the number of progeny it leaves behind.

Charles Darwin had the brilliant insight that natural selection produces new species in essentially the same way that dog fanciers produce new breeds by artificial selection, by allowing only dogs with heritable, desirable characteristics to become the parents of the next generation of dogs. Similarly, natural selection weeds out the less-fit individuals, while permitting the survival of those best adapted to cope with the hazards of their environment or to take advantage of the opportunities it offers. For example, a well-camouflaged creature is not as likely to be noticed by a predator as is a poorly camouflaged one; a butterfly with a long tongue benefits because it can reach the nectar in flowers that are too deep for a butterfly with a shorter tongue. In either case, the favored individuals will probably live longer and leave behind more progeny than the less-favored members of their species.

As natural selection operates generation after generation, useful new heritable traits, no matter how minor, will eventually spread throughout an interbreeding population. If this population becomes geographically or otherwise separated from other populations of its species, it is likely to become a new and different species as more favorable heritable traits—many of them resulting from mutations—accumulate. If some of these newly evolved traits are physiological obstacles or courtship behaviors that preclude reproducing with individuals of other populations of the species, natural selection will have produced a new species that "breeds true" and is reproductively isolated from all other species.

The house fly's life cycle—from maggot to pupa to egg-laying adult—superbly adapts it to its own particular lifestyle. A house fly maggot (larva) is white, soft-skinned, legless, and carrot-shaped. At its broad tail end are two large spiracles. The pointed front end is a greatly modified head that lacks eyes, antennae, and other external structures except for two small, dark, decurved hooks used in feeding. They are all that remains of the complex mouthparts of its distant ancestors. As it grows to a length of as much as a half-inch, the maggot molts its skin twice. When full grown it undergoes its third molt, a very unusual molt to the pupal stage that is virtually unique among the insects except for other flies. When its hormonal system triggers it to undergo this molt, the maggot changes its shape to become ovoid, a cylinder with rounded ends that is only about a quarter of an inch long. Then the soft white skin hardens and darkens. The pupa separates itself from this last and greatly modified larval skin, but remains within it rather than shedding it as other insects do. With admirable parsimony, this last larval skin is not discarded. It has become a puparium that covers and protects the pupa as a caterpillar is protected by its silken cocoon.

After shedding its pupal skin, the adult fly must get out of the hard and sturdy puparium. It accomplishes this in an amazing way, popping off the cap of the puparium with an expansible bladder, almost as large as the head, that can be forcefully everted by blood pressure through an opening on the front of the head. After the adult fly has crawled out of the puparium, it uses the bladder to force its way through any debris—manure, straw, sand—that covers it. When it reaches the surface, the now useless bladder is deflated and retracted back into the head, and the opening through which it was everted closes, leaving a visible scar known as the frontal suture.

House flies may disperse considerable distances from the manure pile or other site in which they developed. In 1916 R. R. Parker of the Montana State Board of Entomology released over 387,000 artificially colored house flies from four points in a small city in Montana and recaptured over 1,000 of them in baited traps that were from 50 yards to about two miles from their release point.[8] He demonstrated that the average fly leads an extremely migratory existence and that a rapid spread over an area of five square miles is fairly common. Later, F. C. Bishopp and E. W. Laake, medical entomologists, found that many of the flies they released were "quite promptly" caught in traps about 1,000 yards from the release point, but some were recaptured much farther away, six miles after less than four hours, and after a longer time as much as 13 miles.[9]

After mating, the recently emerged females lay eggs in almost any decaying waste, but by far prefer animal excrement, especially horse manure. Female house flies produce, on average, a total of about 500 eggs laid in separate batches of about 100 each. The record is held by a captive fly that laid 2,387 eggs in 21 batches. When the temperature is warm, the eggs hatch in about 8 hours but only after 30 hours when it is cool. Given optimal conditions, a whole life cycle—from egg to egg-laying female—can be completed in as little as 10 days. As the economic entomologists Robert L. Metcalf and Robert A. Metcalf write, this fly "combines a large family with one of the shortest life cycles known among insects."[10] In northern areas house flies spend the winter mainly as maggots or pupae.

A male house fly ready to mate seizes a female, sometimes in the air, but the pair descend to the ground before mating. The male caresses the female's head with his front legs, a brief form of foreplay that apparently puts her in the mood to copulate. He then inserts his tubular, rather long intromittent organ, which we can call a penis by analogy, into her genital opening and injects his semen.

The fact that the male house fly has a penis, as do all other insects except for the primitive silverfish and their relatives, is noteworthy, because the penis makes possible internal fertilization. (Insects evolved the penis and internal fertilization hundreds of millions of years before mammals.) The remarkable evolutionary success of the insects would have been impossible if their ancestors had not made the transition from living in the seas to living on dry land, and this transition would have been difficult if not impossible if those ancestors had not evolved the capacity for internal fertilization. In contrast, jellyfish, starfish, oysters, sea anemones, and many other creatures are bound to their life in the sea, because they practice external fertilization—simply releasing eggs and sperm into the water—where they must find each other on their own.

In the house fly and most other insects, the fertilization of an egg proceeds in a more complex fashion than in humans. This complexity gives the female more discretion in the process of fertilization and makes for a more economical use of sperm. A human female receives about 70 million or more sperm at each copulation, but only one of this great swarm of sperm will succeed in fertilizing an egg. All the rest go to waste. Unlike humans, the house fly and almost all other insects have an internal storage pouch for sperm (the spermatheca) that branches from the female's "vagina."

The sperm are stored in this pouch and can be kept alive and healthy for days in the house fly, and in other insects, for weeks, months, or—as in the honeybee queen—for years. The sperm pouch makes it possible for a house fly or almost any other insect to continue laying fertile eggs long after it last mated. As do other insects, the female house fly uses the stored sperm parsimoniously, releasing just a few as each emerging egg moves down the vagina and passes the opening to the spermatheca. Females of a few insect species are maximally miserly, releasing only one sperm for each egg that is to be fertilized. In any case, the sperm enters an egg through a tiny opening in its shell. The ability to store sperm makes it possible for a female to mate at the most propitious time—perhaps late in autumn—but to delay laying eggs until the next spring, when her larval offspring are most likely to survive.

My friend and colleague Tom Frazzetta writes, "Yet even a child can appreciate in a beast a sense of integrated wholeness, of design, of cooperation among its parts."[11] The house fly, the "beast" that is the subject of this chapter, is in our eyes a lowly creature hardly worth noticing. However, like all animals, it is the product of eons of evolution, a beast superbly attuned to its environment, the last in a line of countless ancestors shaped by natural selection. But it will not always be the last in its line. Like its ancestors, it is even now being changed by natural selection—witness how rapidly it became resistant to DDT. Insects evolved a myriad of anatomical, physiological, and behavioral adaptations that help them to survive in complex and unforgiving environments. The fly is no exception; it has many helpful adaptations. I find two to be especially intriguing: the adults' method of feeding and the amazing sensory mechanism that keeps them on course when they fly.

Flies have taste receptors on their feet, a great convenience when searching for food because the feet are the first part of their bodies to touch a surface. If, for example, a fly touches its feet to a leaf wet with honeydew, the sugary excrement of aphids, it knows that it has found food even before extending its long, flexible "snout," the proboscis, which also has taste receptors. That flies can taste with their feet is demonstrated in entomology courses with a simple experiment. Leaving its head and legs unhindered, a fly's wings are attached with a drop of paraffin to a stick as thin as a toothpick. The stick is held in place by a clamp, and if the fly's feet are touched with a brush wet with sugar water, it extends its proboscis by reflex, as if to drink.

Evolution has modified the mouthparts of house flies and other flies from the ancestral chewing type with mandibles to the sponging type, which has no mandibles and can only sip liquids. These sponging mouthparts, which extend downward from the head, are essentially a thick, fleshy proboscis that is elbowed in the middle and ends in a large, expanded organ that consists of a pair of spongelike *labella* closely appressed to each other. The lower surface of each labellum bears many transverse furrows with an inside diameter of only about 0.0005 inch. Within the proboscis is a sucking tube, or food canal, that ends between the labella. Now how does this strange apparatus work? When a fly drinks a thin film of liquid, perhaps from feces outdoors or a slice of meat on your table, it appresses its labella against the film, and the liquid enters the tiny furrows on the labella and flows along them by capillary action to where they all converge between the labella, where the food canal sucks the collected pool of liquid up into the digestive system. Even more amazing is how a house fly eats solid foods such as a dry encrustation of honeydew on a leaf or sugar in a bowl on your table. The fly appresses its labella to the honeydew or sugar and then salivates and regurgitates. When taste receptors on the labella tell the fly that sufficient sugar has dissolved in this liquid, it sponges it up with the labella and sucks it into its digestive system.

As is to be expected, winged insects instantly begin to fly if their feet lose contact with a surface. This is a reflex that is easily demonstrated. Just glue a length of fine thread to the top of the thorax of a house fly, or better, some larger winged insect such as an adult cockroach, so that it does not interfere with the wings. If you lift up a standing fly or some other winged insect by the thread so that it dangles in the air, it will immediately begin to beat its wings and will usually continue to fly until it is exhausted. You can stop it by lowering its feet to a surface or even by letting it grasp a small bit of styrofoam with its feet as it dangles.

All flies, including even the gnats, midges, and mosquitoes, have only one pair of wings. (The scientific name of their order, Diptera, means "two wings" in Greek.) The front wings are present, but all that is left of the hind wings is a pair of tiny, club-shaped organs, the *halteres*, or balancers. A haltere, as Sir Vincent Wigglesworth observes, is "an alternating gyroscope" as opposed to the familiar spinning gyroscope.[12] The thin shaft of a haltere ends at the terminal club, a large, heavy, blood-filled knob. On the base of the haltere's shaft is an array of many tiny sense organs that perceive any slight bending or twisting of the shaft. The entire haltere is a sense organ that stabilizes flight, much as a spinning gyroscope is a "sense organ" that keeps a rocket true to its trajectory. When a fly is on the wing, the halteres rapidly beat up and down. If the fly goes off course by pitching, yawing, or rolling, the halteres are forced out of the plane in which they have been beating. But these gyroscopic sense organs have inertia; in other words, they "want to keep beating in the same plane." As they are forced into a different plane, the shaft of the haltere will undergo just a slight bending or twisting, which is perceived by the sense organs at the base of the shaft, which send a "message" to the central nervous system that tells the fly that it is going off course, enabling it to correct the deviation.

There are many other fascinating adaptations, but they are too numerous to describe in detail here. Among them is a sensitive humidity receptor on the antennae that the fly uses to identify favorable moist resting places. A tiny featherlike appendage on the bulbous last segment of the antenna is a wind speed indicator. The house fly, like other insects, has compound eyes, each composed of 4,000 closely packed, separate, light-sensing elements. Its eyes do not resolve detail as well as do ours, but—as you notice when you try to swat a fly—it is phenomenally adept at sensing movement.

Charles Darwin realized that the framework of his theory of evolution was incomplete because, when he wrote, science knew almost nothing about genetics, the science of the mechanisms of heredity. He wanted to understand how the characteristics of a plant or an animal are passed on from one generation to another, and how the all-important variations—now called mutations—the grist for the mill of natural selection, originate. But when Darwin wrote *The Origin of Species*,[13] virtually nothing was known about this. The beginnings of the modern science of genetics go back to the late nineteenth century, but genetics did not really flower until Thomas Hunt Morgan's experiments with "fruit flies," which he conducted, beginning in 1907, at Columbia University in New York. But more about this in the next chapter.

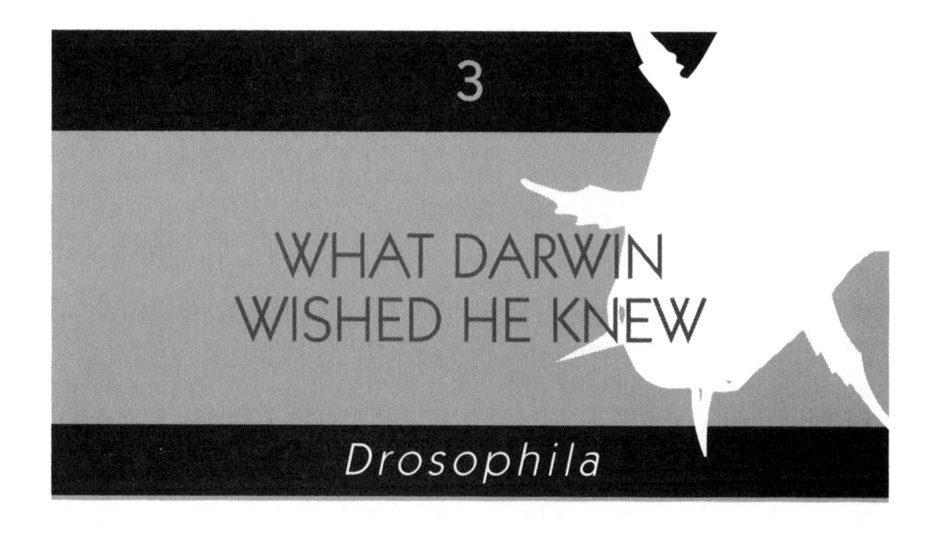

3 WHAT DARWIN WISHED HE KNEW

Drosophila

The "fruit fly" of geneticists, *Drosophila melanogaster,* is technically speaking not really a fruit fly. Entomologists reserve that common name for members of another family of flies, the Tephritidae, which we will consider later and that includes such devastating pests as the apple maggot and the Mediterranean fruit fly. In the larval stage they burrow within and destroy living fruit as it hangs on the tree. This is quite different from the behavior of the species of the family Drosophilidae, correctly known as pomace flies, many of which are attracted only to rotting fruit that has fallen from the tree. To avoid confusion, I will refer to the "fruit fly" of the geneticists by its well-known generic name, *Drosophila*.

The pomace flies constitute a large cosmopolitan family that includes about 2,500 species, in America north of Mexico over 180 species,[1] and in the Hawaiian Islands about 1,000 species generated by an explosive evolutionary radiation (expansion) and found nowhere else in the world.[2] In other words, the first pomace flies to reach the islands from the mainland proliferated into many new species as they took advantage of ecological niches, opportunities for making a "living" not yet exploited by other animals.

The genus *Drosophila*, which includes many species, is only one of several genera in this family. The pomace flies are a tremendously varied lot with many different lifestyles other than the *Drosophilas*' association with fermenting fruit.[3] Some species inhabit rotting leaves and others decaying fungi, some live in fermenting sap, and yet others inhabit feces. *Drosophila*, and probably other pomace flies that inhabit decaying organic matter, eat little or none of the decaying matter itself but rather feed on the yeasts and probably other fungi that grow on it. A few pomace flies are leaf miners that spend their larval stage tunneling and feeding between the upper and lower surfaces of a leaf. A few are scavengers that feed on waste matter in the nests of carpenter bees or mason wasps. Some are predators that eat aphids and scale insects, and finally, a few are parasites that live in the bodies of scale insects, tiny, immobile creatures often covered by a scale of wax.

Male and female *Drosophila melanogaster* come together on rotting fruit. Females seeking a mate release a sex pheromone, a volatile chemical messenger, into the air. Eager males are attracted and woo these females with a brief courtship dance that ends when the male jumps on the female, pushes her wings apart, grasps her body with his legs, makes genital contact, and inseminates her. Although some female insects, such as the handsome cecropia moth, mate only once, *Drosophila* females mate several times during their lifetime of three weeks or more. Multiple matings can dramatically increase the number of fertile eggs that a female insect can lay. A female *Drosophila*'s spermatheca may, just after mating, contain between 500 and 1,000 sperm. With these relatively few sperm she can fertilize, much like a house fly, 500 or more eggs. If she does not mate again her supply of sperm depletes quickly after about six days and she lays only a few more fertilized eggs. Females that mate more frequently—many remate after six days—lay many more fertilized eggs, a thousand or more per female.

Laboratory experiments have shown that female *Drosophila* lay few or no eggs unless they are provided with a diet containing sugar, protein, and vitamins. In nature they obtain these nutrients from rotting fruit and the yeast that causes it to ferment. Warren Spencer noted that *Drosophila*, which means "lover of dew" in Greek, is "hardly a misnomer, for the species [of this genus] without exception seem to require considerable moisture in the environment."[4] According to M. Ashburner and J. N. Thompson Jr.,[5] the eggs, placed just under the surface of the moist food, hatch in the laboratory after two days at a constant temperature of 77° F. The larvae burrow into the food (rotting fruit under natural conditions) and feed almost constantly on the yeasts that grow in it. Four days later, still at 77° F, they are ready to form the puparium in much the same way as do house fly larvae. They first move to a dry place, in a culture bottle to the glass above the wet diet, and in nature to the soil or some other relatively dry site. After four more days, the adult fly emerges from the pupa within the puparium and forces its way out just as do house fly larvae.

Drosophila can be pests. They are often found in homes if overripe fruit is there to attract them, a problem easily avoided by disposing of spoiling fruit. *Drosophila* in the home are a minor problem, but they can be extremely troublesome at canneries where fruit is being packed or at processing plants where ketchup is produced. For example, tomatoes harvested in the field and brought to processing plants by the truckload are likely to crack or even be crushed. They attract swarms of *Drosophila*, some of which inevitably end up in the finished product. Most of the foods we eat contain small quantities of insects or insect fragments, because it is impossible to totally exclude them as a food is grown, harvested, shipped, and processed. It might be possible to produce foods that contain no insects or insect fragments, but that would require such meticulous care that the price of these foods would be astronomical. Recognizing this, the US Food and Drug Administration has set a maximum legal limit on how much insect contamination is permitted in a food.[6] For example, 3.5 ounces of peanut butter can legally contain up to 30 insect fragments, the same quantity of frozen broccoli up to 60 aphids (plant lice) or other tiny insects, and tomato sauces (presumably including ketchup) up to 30 fly eggs, 15 fly eggs and one maggot, or two maggots.

During Darwin's time, in a climate of genetic ignorance, some of his critics argued that natural selection would be impossible because of the then common but erroneous belief in blending inheritance. They thought that an advantageous variation would be diluted out of existence when the variant animal and its descendants mated with normal individuals. Thus, they thought, if a butterfly with a freakish but advantageous 4-inch-long tongue mated with an individual with a normal 2-inch tongue, all their progeny would have 3-inch tongues; their grandchildren would have tongues only 2.5 inches long, and their descendants would have even shorter tongues. In this way, they theorized, the advantage of having a long tongue capable of sipping nectar from deeper flowers—or any other advantageous variation for that matter—would very soon be lost.

In *The Variation of Animals and Plants under Domestication*, published in 1868, Darwin tries to counter the arguments of the believers in blending inheritance by proposing the theory of pangenesis as the mechanism of heredity.[7] It was a valiant attempt but doomed to failure because he and virtually no one else then knew of the 1866 article by Gregor Mendel, which described his experiments on inheritance in garden peas, and which would become the basis of modern genetics and thereby ultimately bolster the theory of evolution.[8] But it was not until 1900 that Mendel's work, published in an obscure journal, finally came to the notice of science. In essence, Darwin's pangenesis hypothesis[9] proposed that every cell of a plant or animal produces extremely minute "gemmules"—the units of inheritance—that "are collected from all parts of the system," are passed to the sex organs, and then on to the next generation after they have been incorporated in the eggs and sperm. He credited variations to "the reproductive organs being injuriously affected by changed conditions, and in this case the gemmules derived from the various parts of the body are probably aggregated in an irregular manner, some superfluous and others deficient." Monroe Strickberger writes in his recent textbook that by 1871, Francis Galton, Darwin's cousin, had already disproved the pangenesis hypothesis to his own satisfaction when he found that transfusing their blood, which presumably contained gemmules, among different strains of rabbits did not affect their heredity.[10] However, Darwin had never said that gemmules are carried in the blood.[11]

Not long after Darwin's time, the science of genetics began to progress on two fronts: studies of the structure of cells, which ultimately led to the discovery of chromosomes, which are the bearers of the material of heredity; and experiments with the hybridization of plants and animals, studies of genetics in action, which demonstrated that heredity is not a blending process—rather, characteristics are inherited by means of discrete units on the chromosomes that are not diluted by the presence of other units of inheritance.[12]

Many microscopists contributed to our understanding of cells, notably by discovering the cell nucleus, which contains the chromosomes, and the nucleus's role in the creation of new cells by the division of existing cells. As my friend and colleague David Nanney pointed out to me, "genetics is in part a technology-driven science." During the golden age of microscopy, about 1875 to 1900, the microscope was greatly improved: increasing magnification enhanced the clarity of the image, and thereby made it possible to see ever-finer details. The recently invented microtome cut microscopically thin sections of tissue in much the same way that the slicing machine in a delicatessen slices a sausage. Fixed to a glass slide, these sections were stained with a variety of newly developed dyes that enormously enhanced the visibility of cells and the nucleus and other organelles within them. In 1882 Walther Flemming,[13] with the aid of dyes and fixatives, saw strongly stained bodies in the nucleus and called them *chromosomes*, "colored bodies" in Greek. They formed during cell division, doubled in number by splitting, clustered into separate groups of equal numbers, and after cell division each of the daughter cells had one of the two clusters. In 1892 August Weismann,[14] building on Flemming's work, conceived the germ plasm theory of inheritance, which disposed of the need for postulating the existence of "gemmules" by proposing that, as we now know with certainty, chromosomes bear the material of heredity, and that they have all the hereditary elements, now called *genes*, that are needed to produce a whole new individual.

In 1900 Hugo De Vries (Dutch), Carl Correns (German), and Erich von Tschermak-Seysenegg (Austrian), discovered, independently of each other, Mendel's long-overlooked article on garden peas.[15] They had previously done similar experiments with several different kinds of plants, confirming Mendel's discovery that genes are discrete entities that are transmitted independently of each other. In other words, they showed that inheritance is particulate rather than blending.

The garden pea is a good subject for genetic experimentation, because individual plants vary in many characteristics, such as seed color, seed texture, flower color, and plant height. Mendel variously crossbred these different varieties, but one of these crosses is enough to show us that inheritance is indeed particulate. He crossed varieties with smooth or wrinkled seeds, characteristics controlled by alternative forms—alleles—of the same gene. The allele for smoothness (S) is dominant over the recessive allele for being wrinkled (s). A smooth seed may, thus, be of two genetic types, one with two dominant alleles for smoothness (SS) and another with a dominant allele (S) "overwhelming" a recessive allele for being wrinkled (Ss). Particulate inheritance is shown by the seed types of the progeny and grandprogeny of an SS and an ss parent. Their progeny all have smooth seeds, but all are Ss, as revealed in the next generation, the grandchildren. When two Ss parents produce offspring, 25 percent of them have wrinkly seeds (ss) and 75 percent have smooth seeds, a third of the latter SS and two-thirds Ss.

The specter of blending inheritance, an impediment to the acceptance and further development of the theory of evolution, had now been exorcised by the beginnings of the modern science of genetics. The importance of this cannot be overstated. Evolution has become the central and unifying concept of all biology, more so than ever since molecular biologists have shown that DNA is the stuff of the genes, genes that control or at least influence everything from the anatomy to the behavior of an organism. As genetics continued to develop, *Drosophila* soon became the geneticists' laboratory animal of choice, and made possible the all-important work of Thomas Hunt Morgan in his laboratory at Columbia University. David Nanney told me that this tiny and seemingly insignificant insect was then and is now a giant—perhaps *the* giant—among the animals used in laboratory research. Biologists have used organisms domesticated for other purposes, such as peas, corn, and cattle, in scientific studies, but *Drosophila* was the first organism deliberately domesticated as a subject for laboratory research.

Drosophila has three major advantages as a laboratory animal. First, as described by Warren Spencer,[16] in nature these flies can be trapped by the hundreds with fermenting mashed banana, and can then be used to start a laboratory colony that can be maintained for many years: hundreds of *Drosophila* generations, in half-pint milk bottles plugged with cotton and containing any one of several diets. At first, researchers used only mashed banana, and even today one of the commonly used diets consists of mashed banana, yeast, agar gel, water, corn syrup, barley malt, and small quantities of preservatives. What is particularly convenient for researchers is that *Drosophila* completes a generation in only 10 days. In the one year that Mendel waited for the completion of a generation of peas, 36 generations of *Drosophila*—36 opportunities for experimentation—could have been created. Finally, because they are small, about one-quarter the length of a house fly, a colony requires a minimum of space. Consequently, *Drosophila* can be made available in very large numbers, by the hundreds or even thousands, making it possible to do experiments that require the use of many individuals.

By the end of the nineteenth century, there were two different views on the nature of the variations that are the raw material for natural selection. One school, the selectionists, agreed with Darwin that natural selection acting on small continuous variations is the main process controlling evolution. The opposing school, the saltationists, wrongly interpreting Mendelian genetics, argued that natural selection can act only on large, discontinuous variations, major mutations that cause abrupt changes between species. To take an extreme example, insects might acquire wings in one big step, rather than evolving them gradually, beginning, perhaps, with the ability to glide from one tree to another.

These opposing views were reconciled by what has come to be called the neo-Darwinian (or Modern) synthesis. Morgan, arguably the greatest of the early geneticists and who made *Drosophila* the most important research animal in genetics, discovered much of the knowledge that eventually made the neo-Darwinian synthesis possible. At first Morgan was highly skeptical of the theory of evolution, but as his onetime assistant, Herman Muller, wrote in his 1946 obituary of Morgan, "the end result of the skepticism, since it was combined with experimentation and exact observation, was to lead some of this generation, and most of the next, to a vindication of the Darwinian essentials after all, and to an effective implementation of the Darwinian theory which joined it up with a scientific view of living matter in general."[17] Morgan's work demonstrated that chromosomes bear the factors of heredity, that genes—although invisible—can be shown to exist. "Without Morgan [and *Drosophila*]," wrote Ronald Clark in his biography of Darwin, "there might . . . have been no neo-Darwinism that successfully combined the theory of natural selection with the discovery of Mendelian genetics."[18] In 1933 Morgan won a Nobel Prize for his work in genetics.

The neo-Darwinists realized that natural selection is best understood by focusing on populations of organisms rather than on only the individual. A population forms a huge "gene pool"—with different genes present in different frequencies—from which individuals of each generation, via their parents, draw their personal combinations of genes, often at random. When these individuals enter their ecosystems, natural selection acts on their bodies and behaviors, which are determined largely by their particular genetic blueprints, thereby changing the frequency of some genes in the gene pool and having no effect on others. According to the neo-Darwinian synthesis, evolution is an ongoing process in which randomly occurring mutations introduce new genes, whose frequencies in the gene pool change through time due mainly to natural selection, but also by other factors such as migration or even "genetic drift," evolutionary change caused by random chance.

Morgan and his colleagues and students worked in a small room, 16 × 23 feet, which came to be known as the "fly room," and into which they crowded eight desks and their colony of *Drosophila*, which fortunately did not take up much space. Morgan's group began their research in 1907, but it was not until 1910 that Morgan discovered his first mutant, a male *Drosophila* that had white rather than the usual red eyes. When he crossed the white-eyed male with a normal red-eyed female, their progeny were all red-eyed, and when he crossed the progeny with each other, one-quarter of their offspring were white-eyed, but three-quarters were red-eyed. This confirmed the results of Mendel's experiment with smooth- and rough-skinned peas, but there was an important difference: all of the white-eyed flies in his experiment were males. It was already known that female organisms contain two X chromosomes and that males contain one X chromosome and one Y chromosome. Morgan did a series of experiments that for the first time demonstrated sex-linked inheritance, that the recessive gene for white eyes is carried by females but expressed mainly in males, although white eyes occasionally occur in females with an unusual genetic makeup. There are similar but not identical cases of sex-linked inheritance in humans. For example, red-green color blindness, which occurs in about 8 percent of white males but almost never in females, and hemophilia, which was passed on from Queen Victoria of England to males of several of the royal families of Europe.

The white-eyed fly became justly famous and, according to Ronald W. Clark,[19] legend has it that when Morgan visited his wife in the hospital, where she had just given birth to a daughter, she asked him, "Well how is the white-eyed fly?" It is said that Morgan replied enthusiastically at some length before he stopped abruptly and finally asked, "And how is the baby?"

Morgan soon discovered more mutations in *Drosophila*. If flies with genes for two different mutations were mated, the two mutations were often passed on together from one generation to the next because both genes were on the same chromosome. But Morgan found that this did not always happen. He postulated that this occurred because the part of a chromosome carrying one of the genes had changed places with a part of another chromosome, leaving the other gene in place on the original chromosome. As geneticists put it, crossing over had occurred. He also observed that some pairs of genes were much more likely to be passed on together than were others. He reasoned that the ones that were likely to remain together were close together on the chromosome, making it unlikely that in crossing over the chromosome would break between them. The discovery of this phenomenon, called linkage, ultimately made it possible to determine the distance between genes and thereby to map the positions of the genes on a chromosome. Muller wrote that "Morgan's evidence for crossing over and his suggestion that genes further apart cross over more frequently was a thunderclap, hardly second to the discovery of Mendelism, which ushered in that storm that has given nourishment to all of our modern genetics."[20]

The discoveries of Morgan and the other early geneticists gave us the context within which the importance of DNA, the very material of the genes, can be understood. We now know that a chromosome is a very long length of DNA, a double helix of spiraling strands, folded and refolded upon itself. The genes are spaced along the length of the double helix and each is responsible for the formation of a particular enzyme or other protein. Molecular biologists now have determined the complete DNA complement of many organisms such as bacteria, plants, and several animals including a mosquito, *Drosophila*, and even humans. Many of these genes, responsible for various anatomical, physiological, and behavioral aspects of an organism's life, have been identified and isolated. Within the past two decades, molecular biology has made amazing progress; indeed, molecular biologists can now remove genes, add genes, or even transfer genes between different species of organisms to create genetically modified crops or other organisms (GMOs).

The development of resistance to DDT by the common house fly is by no means the only well-documented example of natural selection in action. There are many, for example, within the last twenty years, as you will see on the following pages: two closely related species of beetles that feed on the roots of corn plants evolved ways to overcome a common agriculture practice, crop rotation, that since the nineteenth century had been a virtually perfect noninsecticidal control of these two destructive pests. The most intriguing aspect of this development is that each of these beetles has evolved its own unique way, completely different from the other, of circumventing crop rotation. This is very interesting but does not come as a complete surprise, for now we know that probably all organisms have in their genetic makeup a store of many latent mutations that, when the need arises, can be called into service by natural selection.

4

NATURAL SELECTION OUTFLANKS FARMERS

Corn Rootworms

Since he began growing corn and soybeans in central Illinois several decades ago, a farmer—I will call him Matt—has been rotating these two crops in his fields, never growing corn in the same field for two years in a row, always following corn with soybeans. This has been a wise move, because it prevents root-feeding corn rootworms, which will not attack soybeans, from damaging his corn plants, thereby obviating the need for a costly application of an insecticide. On a day in July of 1998, Matt checked one of his cornfields, and instead of neat rows of erect green plants he saw a tangle of wilting stalks leaning in all directions. His crop had been devastated. Matt pulled up a few plants and saw that most of the roots had been destroyed, making it impossible for the plants to stand erect. This was clearly the work of root-feeding corn rootworms, the larvae, or grubs, of two closely related species of the leaf beetle family. Matt wondered how this could have happened when he had, as always, rotated corn and soybeans.

Crop rotation effectively controlled corn rootworms for more than a hundred years. Why it worked will become apparent when we understand the feeding habits and the life cycles of these two beetles. Why it ultimately failed can be explained only by Charles Darwin's concept of natural selection.

The northern corn rootworm (*Diabrotica barberi*) was the only corn rootworm in Illinois until 1964, when the western corn rootworm (*Diabrotica virgifera*) invaded from the west and eventually became far more abundant than the northern corn rootworm. The life history and habits of these two species are very similar—at least they were before crop rotation failed to control them. The original scenario began as follows: in late August and early September the eggs are deposited a few inches down in the soil at the base of a living corn plant—and almost never anywhere else. The eggs are in diapause (a winter sleep). In spring, diapause is terminated and the eggs hatch. The tiny grubs, which can move through the soil for only a very short distance—certainly not from one field to another—feed on the roots of a young corn plant, attracted to them mainly by the carbon dioxide they emit. With the exception of certain prairie grasses, they will feed only on corn, and will die rather than eat the roots of other plants, such as soybeans. After a month or more of greedy feeding, the grubs are full grown—most by early July—and move away from the roots to pupate in a cell they form in the soil. In July and August the adult beetles emerge from the soil and feed on corn silks, the pollen of corn, and the flowering parts of many other plants. At this time females of both species emit a sex pheromone that attracts males of their own species. After mating, the eggs are laid. The adult beetles are killed by the first severe frost, and in the winter both species are represented only by eggs.

Although the two species of corn rootworms now have similar biological associations with corn and have similar adverse and sometimes disastrous effects on this plant, they came to be associated with corn in different ways and in widely separated geographic areas. There is little doubt that western corn rootworms first used corn as a food plant in Meso-America, probably Mexico, perhaps as much as 4,000 years ago when, according to Paul Mangelsdorf, Native Americans discovered, domesticated, and grew corn as a staple food.[1] It may well be that, even earlier, teosinte, the ancestor of corn, or other wild perennial grasses were on the western corn rootworm's menu. These insects eventually became dedicated eaters of corn and followed this crop northward as the cultivation of corn by Native Americans spread across what is now the United States. This northward expansion of the western corn rootworm probably started over 3,000 years ago, when the native people of what is now northern Mexico and the southwestern United States began growing corn as a crop. Today, western corn rootworms have spread throughout much of the northern part of the corn belt of North America.

The association between corn and northern corn rootworms came about much later and in a different way. These insects also originated in Central America but invaded the tall grass prairie of the midwestern United States long before corn had been domesticated by Native Americans. As postulated by F. M. Webster,[2] they probably had no contact with large plantings of corn until some time after 1865, when European settlers began to grow this crop on the prairies of the Midwest. Then northern corn rootworms made a massive switch to corn from their original food plants, perennial prairie grasses such as the big blue stem, the tall grass that came up to the waist of a man on a horse. Terry Branson and James Krysan[3] believe this happened in an area centering on eastern Missouri and central Illinois. Since then northern corn rootworms have spread east and west so that they now can be found throughout the corn belt.

The behaviors that enabled northern and western corn rootworms to exploit wild perennial grasses predisposed them to similarly exploit corn, which is, after all, a species of grass. One obstacle in the way of these rootworms' switch from perennial wild grasses to corn is that corn is an annual rather than a perennial. This obstacle was largely obviated by a common practice of some corn growers: planting corn in the same field year after year. Although corn is indisputably an annual, this practice makes it, for all practical purposes, a perennial, available to newly hatched corn rootworm larvae spring after spring.

Western corn rootworms may or may not be infested with a bacterium of the genus *Wolbachia*, the first known species of which was found in the reproductive organs of mosquitoes collected early in the twentieth century in a storm sewer near the Harvard University Medical School. The species that infects the western corn rootworm is transmitted from beetle to beetle in the eggs of females but not in the sperm of males. The effect of *Wolbachia* on the geographical distribution of this insect, recently elucidated by Rosanna Giordano and her coauthors,[4] is economically significant and truly amazing. The western corn rootworms now established in the northern part of the corn-growing areas in the United States and in southern Canada are infected with *Wolbachia*. An adjoining population of the same species in Mexico and parts of the southwestern United States is not infected. When uninfected males of the latter population mate with infected females of the northern population, the usual number of normal progeny results. On the other hand, when infected males of the northern population mate with uninfected females of the southwestern population, very few or no surviving offspring are produced. Giordano demonstrated that the *Wolbachia* bacteria are, as the observed results suggest, responsible for this strange pattern of reproductive success and failure. Why are matings between infected males and uninfected females rarely successful? An infected male transmits no *Wolbachia* to his uninfected partner, and when his sperm enter her eggs, the eggs seldom develop normally. Sperm that develop and mature in the body of an infected male are somehow altered by the *Wolbachia*—as yet no one knows how—so that they cannot properly fertilize eggs of uninfected females, although they can successfully fertilize eggs of infected females. An experiment done by Giordano and her coauthors substantiates this. They found that matings between northern females freed of *Wolbachia* with the antibiotic tetracycline and uninfected males from the Southwest were fully fertile and produced the normal number of offspring. As Giordano pointed out to me, it is to *Wolbachia*'s evolutionary advantage to avoid diluting its population by preventing uninfected females from reproducing. From an economic point of view, *Wolbachia* may well be benefiting agriculture by preventing the southwestern strain of the western corn rootworm from invading the southern United States.

After entomologists came to understand the life history, habits, and ecology of the northern and western corn rootworms, especially their near total reliance on corn as their food plant, it became obvious that the damage they inflict on corn could be avoided by rotating crops—by not planting corn in the same field for two or more successive years. (This strategy was first suggested in the nineteenth century by one of the great early entomologists, Stephen Alfred Forbes, state entomologist of Illinois.) Because these beetles have only one generation per year and die after they lay their eggs in late summer and fall, they will be represented only by eggs in winter, all of which will be in the soil of fields in which corn grew that year. The larvae that hatch from the eggs the following spring will die of starvation unless the roots of a living corn plant are nearby. It follows that the soil in a field in which some other crop had grown the previous year, but is now planted with corn—known as first-year corn—will contain no corn rootworm eggs and the corn will not be infested.

In the nineteenth and early twentieth centuries, corn was grown in rotation with hay, clover, alfalfa, or small grains. In the 1920s soybeans began to replace these other crops, and the rotation became more and more an alternation between corn and soybeans. By the early 1990s in Illinois, only 27 percent of the corn crop was planted in fields that had previously been in corn, but almost 67 percent was grown in rotation with soybeans and another 6 percent in rotation with other crops.[5]

Crop rotation prevented corn rootworm infestations for over a hundred years and continues to do so in many areas of the country. But all organisms are constantly subjected to natural selection, and the two species of corn rootworms have now evolved strains that can, each in its own way, get around crop rotation. The northern corn rootworm was the first to find a way to circumvent crop rotation. Instead of terminating diapause after the first winter and thus hatching when only soybeans or some other unacceptable crop is available, many of the eggs now remain in diapause longer and do not hatch until the spring after the second winter, when corn is once again available in a rotation of corn with soybeans or another crop. Entomologists refer to this phenomenon as extended diapause. Only a few scattered signs of extended diapause were seen during the first half of the twentieth century, but by the 1980s and 1990s extended diapause had become widespread and had resulted in losses to first-year corn in North Dakota, South Dakota, Nebraska, Minnesota, Iowa, Illinois, and Michigan. Northern corn rootworms with an extended diapause are spreading and will eventually be prevalent throughout the corn belt.

In 1962 H. C. Chiang reported that of 2 of 676 northern corn rootworm eggs (slightly less than 0.3 percent) collected in Minnesota continued in diapause through a second winter. But that soon changed.[6] By 1984 James Krysan and his colleagues reported that 40 percent of 329 eggs laid by northern corn rootworm beetles collected in nearby South Dakota had an extended diapause.[7] In 1992 Eli Levine and his colleagues[8] found that of 311 eggs laid by northern corn rootworm beetles collected at the same site in South Dakota, over 50 percent of those that hatched had an extended diapause: 20.6 percent hatched after two winters, 20.9 percent after three winters, and 9.6 after four winters. They found that in Illinois a similar proportion of the eggs of the northern corn rootworm had an extended diapause: of 777 that hatched, 42.1 percent, 8.0 percent, and 0.3 percent, respectively, did not hatch until after two, three, or four winters had passed.

Western corn rootworms found an entirely different way to circumvent crop rotation, but not until several years later. As Eli Levine and his colleagues[9] reported, the first indication that western corn rootworms had found a way to get around crop rotation was seen in 1987 near Piper City in east-central Illinois. Ever since then this insect's ability to circumvent crop rotation has been spreading from this focal point—by 1995, to 23 counties in nearby Illinois and Indiana. By 1997 rotation-circumventing western corn rootworms, presumably carried by westerly winds, were rapidly spreading eastward and had been found as far away as southern Michigan and northwestern Ohio. It will not be long before they reach the corn-growing areas of Delaware and Maryland. Their spread in other directions, which is not aided by the prevailing westerly winds, has been much slower.

Rotation-circumventing western corn rootworms synchronize with rotated corn (first-year corn) by laying many of their eggs in soybean fields rather than in cornfields. They seem to have no special preference for laying their eggs in soybeans. They lay them in association with several other plants, but most of their eggs will, of course, be laid in soybean fields because soybeans are by far the most abundant plants other than corn in most midwestern farming areas. Western corn rootworm larvae do not survive on a diet of soybean roots and although adult western corn rootworms will eat soybean leaves, they do not survive if they have nothing else to eat. If corn and soybeans were not rotated, but planted in the same field year after year, the western corn rootworm's new behavior of laying its eggs in soybean fields would be suicidal, because the larvae that hatch from these eggs would not coincide with corn, which is practically their only food. In a yearly corn-soybean rotation, larvae that hatch after the first winter from eggs laid in soybean fields will coincide with the corn that will be planted the next year.

It is truly remarkable that northern and western corn rootworms have through natural selection (explained above) evolved two totally different ways of circumventing crop rotation and thus coinciding with their food plants. For a hundred years farmers who rotated corn with other crops avoided the depredations of corn rootworms, but northern and western corn rootworms have now found, through natural selection, two ways—a different way for each species—of circumventing crop rotation and coinciding with corn—thus avoiding what would for them be a catastrophe.

Darwin would have been dismayed by the farmers' plight, but he would have been delighted by this amazing example of natural selection in action. Some people might object to my use of the term *natural*, arguing that in these cases selection is artificial, because the crops are rotated by people rather than by "nature." You may use whichever term pleases you. But the fact remains that selection is occurring, and we are now seeing evolution in action, just as we did with the house fly.

Natural selection can alter the anatomy, physiology, or behavior of a population of animals, but this population can become a separate species only if it is reproductively isolated from the other members of its species. That is to say, it cannot be considered a true species if it interbreeds with any other species to any significant extent. Most biologists think that reproductive isolation can be effected only by geographic isolation, as when two populations of the same species are separated by an impassable barrier such as a mountain range or a large body of water. Others argue that speciation can occur without geographic separation—if, for instance, two populations living in the same place become isolated from each other because they feed, mate, and live on two different species of plants. Such seems to be the case with the apple maggot, one of the true fruit flies you will meet next.

5

HOW A SPECIES BECOMES TWO SPECIES

Fruit Flies

In a 1922 US Department of Agriculture Farmer's Bulletin, A. L. Quaintance and E. H. Siegler said of the apple maggot: "The work of this insect in its earlier larval stages is often very deceptive, and apples that appear sound externally are frequently infested with one or more maggots. As soon as the infested apples become mellow, however, the maggots develop rapidly and can be readily detected by the brownish tunnels which are often visible through the skin, especially with varieties having light or yellowish colored skin. . . . The larvae or maggots make winding burrows or tracks throughout the flesh of the fruit and often reduce it to a brown pomace-like consistency, rendering it absolutely worthless for market. . . ."[1]

The apple maggot is one of the true fruit flies, a member of the family Tephritidae, not to be confused with the family Drosophilidae, the pomace flies, often incorrectly called fruit flies. Pomace flies lay their eggs on the surface of a rotting fruit, but fruit fly maggots that feed on fruit—not all of them do—live and feed within healthy fruits still hanging on the tree. Fruit flies have a sharp, sturdy ovipositor (egg-depositing organ) with which they pierce the fruit and place an egg in its inner flesh—the tissue that the maggots will eat—and in which they will be protected from desiccation and predators.

There are about 4,200 known species of fruit flies, but only a few, about 40 species, can be reasonably designated as pests. Most of the known innocuous species will feed on only one species of plant or a few closely related ones. But to the contrary, most of the pest species are highly polyphagous, that is, larvae can survive and grow in the fruits of dozens, scores, or even hundreds of different kinds of wild or cultivated fruits.

The host plant preferences of the apple maggot (*Rhagoletis pomonella*) are an important consideration in an ongoing friendly debate about the circumstances under which natural selection can "create" new species. The predominant opinion is that speciation is possible only if two populations of a species come to be separated by a geographic barrier, such as an ocean or a mountain range, that precludes interbreeding between them. Only then, it is argued, can the two populations accumulate enough genetic mutations that control physiology and behavior, particularly sexual behavior, to reproductively isolate them from all other species, the sine qua non for the existence of a "true" species. The other opinion—and this is where the apple maggot comes in—is that speciation can occur without geographic isolation if two populations, even if they live in close proximity, are separated by a behavioral difference such as utilizing different host plants.

The apple maggot is a native of North America whose original hosts, hawthorns, are in the same family as the apple and bear small, applelike fruits. Apples were introduced into North America early in colonial times, but it was not until 1866, over 200 years later, that *pomonella* was found infesting apples. There are now two "host races" (as evolutionary biologists put it) of this fly, one that infests hawthorns and another that infests apples. They look the same, but behavioral differences between the two races tell us that they are at least on the way to becoming separate species, but it is not clear that they have made it all the way. The apple race terminates diapause and emerges from the soil as an adult earlier in the season than does the hawthorn race. This seasonal difference in emergence makes it unlikely that the two host races will interbreed. When the apple-host race emerges, hawthorn fruits are not yet available. There is no way to find out if the apple race would prefer apple to hawthorn if both were available in nature at the same time. However, in the laboratory, females of the apple race are more willing to lay eggs in apples than are females of the hawthorn race. Because these and most other fruit flies mate on the host plant, the preference for one plant or the other tends to be perpetuated, because an individual is most likely to mate with an individual of its own host-plant race.

A world authority on fruit flies, Stewart Berlocher of the University of Illinois Department of Entomology, told me that close to half of the fruit flies are a diverse lot that do not feed on fruit. The larvae of one large group feed on plants of the daisy family, most in the flower heads but a few form galls (tumorlike swellings) on the stems, roots, or flowers. Maggots of a species of the genus *Plioriocepta* burrow in the young shoots of asparagus, which egg-laying adults locate by sight, responding to their vertical aspect. The species of the genus *Blepharoneura* feed only on plants of the squash family. They are generally strictly host specific, feeding on only one species of squash. Many are even choosy about what part of the plant they utilize. The larvae of some feed only in male flowers, others only in female flowers, some only in seeds or fruits, and a few only in stems. He also told me that in Indonesia, New Guinea, and parts of Australia there are fruit flies that feed on rotting wood. Some, species of the genus *Phytalmia* and a few related genera, are among the most anatomically bizarre of all the true flies. These males have huge, spreading "antlers," projections of the side of the head, which they use, like male deer or elk, in pushing matches with males that intrude on their territories, sites on rotting logs to which females, potential mates, come to lay eggs. Some minute larvae of the genus *Euleia* "mine" in the leaves of plants of the carrot family, tunneling in the soft tissues between the upper and lower surface layers of a leaf. Larvae of the aptly named Australian species *Termitorioxa termitoxena* live in tunnels that termites make in living trees, feeding on the evil-smelling liquid that fills much of the tunnels. There is even a predator among the fruit flies. *Euphranta toxoneura* larvae invade galls (abnormal growths on plants), and eat their rightful owner, the sawfly larva that caused the gall to grow.

In April of 1929 an incipient catastrophe, an invasion of foreign fruit flies, threatened the orchards of Florida. Maggots of the Mediterranean fruit fly were discovered in a locally grown grapefruit. This insect, nicknamed the "medfly," is one of the world's most destructive pests, infesting the fruits of at least 253 species of plants, among them many wild ones and most if not all the important cultivated species. Among the latter are oranges, peaches, plums, cherries, pears, apples, figs, avocados, bell peppers, tomatoes, and even the fleshy outer rind that covers walnuts, to name a few. The medfly (*Ceratitis capitata*) is a native of South Africa that commerce has spread to tropical and subtropical areas all over the world. The infestation in Florida was its first appearance in America north of Mexico, but since then it has been introduced and reintroduced several times in Florida and California. Each time it has been eradicated, although James Carey makes a plausible case that a very small permanent population is holding on in California.[2]

In 1929, 15,000 square miles of Florida were quarantined with road blocks enforced by National Guard troops to prevent the further spread of the medfly. No fruit of any kind was allowed to leave the quarantined area. Because no suitable insecticide was then available, the flies were eradicated by picking and destroying every wild and cultivated fruit in infested areas. This required an army of about 6,000 employees, took 18 months, and cost about $7 million (74 million in current dollars)—a bargain if you consider that the yearly loss to this insect would have been at least three times as much, and that Florida was free of this pest for 27 years, until it reappeared in 1956 and was again eradicated. Today medflies are eradicated or contained by a combination of insecticides and the amazing sterile male method, which will be described in detail in the last chapter of this book.

The fact that the apple maggot is one of the principal subjects of the evolutionary controversy about the way in which species evolve has sparked detailed studies of its mating behavior and choice of food plants. Furthermore, because of its status as an important pest, it has been the subject of many other detailed studies of its natural history, ranging from how a female finds and recognizes an apple in which to lay an egg to the way in which she is assured that her offspring will not be impacted by competition from other larvae of their own species.

One of the goals of this research is to find noninsecticidal alternatives to control the apple maggot, which is vital because apples, attacked by a complex of numerous pests, are drenched with many applications of expensive and environmentally undesirable insecticides. Furthermore, because apple pests are rapidly developing resistance to these insecticides, orchardists are on a treadmill, which cannot continue indefinitely, of replacing failed insecticides with new ones.

Except in the southern part of its range, the apple maggot has only one generation per year. By late August and early September, the larvae are full grown and ready to burrow into the soil beneath the tree, where they will form the puparium that encloses the pupa (in the same way that house flies do). Stewart Berlocher told me that a few larvae drop to the ground after boring out of a fruit still hanging on a tree, but that most fall to the ground with a fruit, which may drop from the tree prematurely if it is infested with apple maggots. The larvae are induced to go into a pupal diapause by the short days and cool temperatures of the end of summer. Diapause is terminated and the adults emerge the following summer, but only after the pupae have experienced the cold of winter. In the laboratory diapause can be terminated by holding puparia in a refrigerator for several weeks.

Adult apple maggots, which are just a little smaller than house flies, emerge from the soil in June and July, but do not reach sexual maturity and lay eggs until about 10 days later. In the meantime they disperse and search for foods that provide sugars as a source of energy and the proteins or amino acids required for reproduction. Among these foods, listed by E. F. Boller and Ronald Prokopy, are honeydew, plant exudates, yeasts, and the liquids on insect feces and bird droppings.[3]

According to Ronald Prokopy and Jorge Hendrichs, as sexual maturity nears, the flies congregate on the host plants of the larvae, where they mate and lay their eggs.[4] Other fruit flies, such as the pestiferous oriental fruit fly and many of its relatives, mate at dusk, but apple maggot adults and at least some of their relatives will mate at any time during the daylight hours as long as it is warm enough. The mating behavior of the males is attuned to the reproductive cycle of the females. Early in the season, before females have started laying eggs, mating occurs on foliage. Males, sometimes in groups, perch on the upper side of a leaf and release a sex attractant pheromone that attracts females. At this time females are apparently eager to mate, and the males—at least most of them—openly approach a female face to face. Later in the season males switch their attention to the fruits, where most of the females are then to be found. A male claims a fruit as his personal territory and defends it against intruding males, "boxing" with his front legs as he and the intruder stand on their hind legs. Females that come to a fruit to lay eggs are much less receptive to males than they had been. They really have no need of them. They already have a store of sperm from a previous mating in the sperm pouch and they are too busy with the chores of motherhood to put up

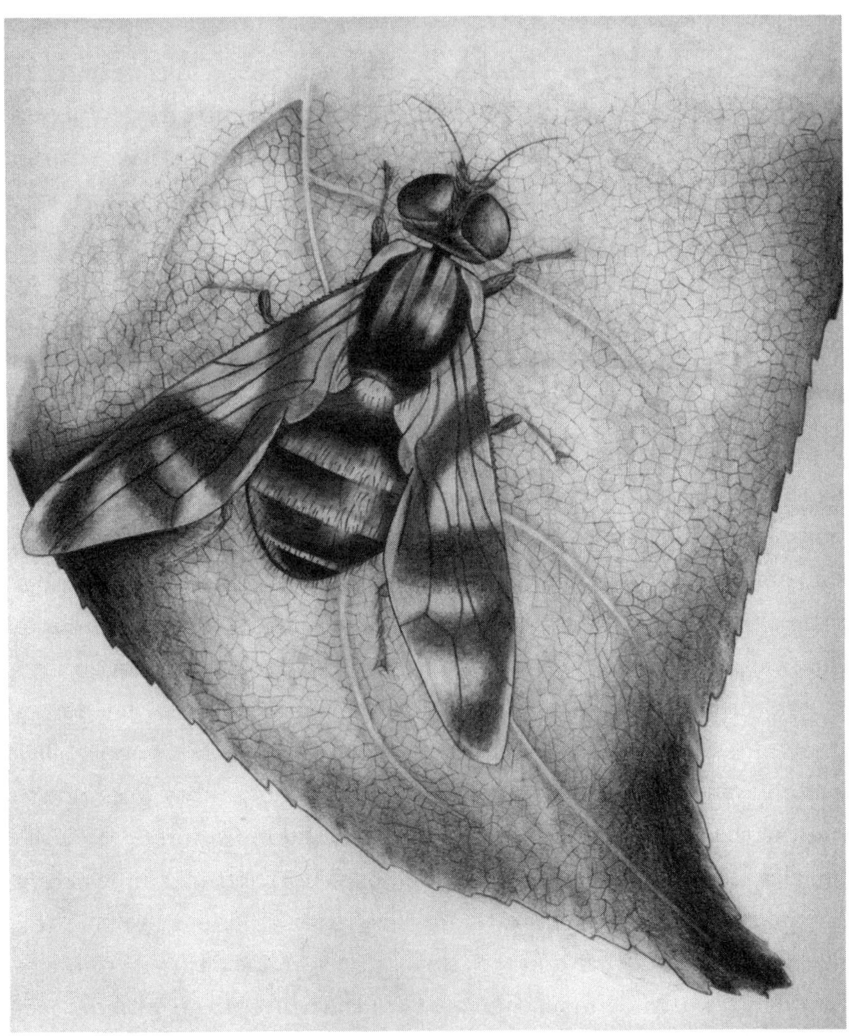

An adult apple maggot perched at the tip of an apple leaf.

with importuning males and usually try to discourage them. Consequently, natural selection has changed the way in which a male approaches one of these busy females. Determined to fertilize her eggs, instead of courting her openly, face to face, he approaches her from behind, and if she is preoccupied with preparing to lay an egg, he often manages to mount her and force copulation. Inseminating a female who is going from fruit to fruit to lay her eggs is especially advantageous. When his sperm enter her spermatheca, they will cover most of those of her previous mate. Because, as in most insects, the last sperm to enter the spermatheca are generally the first to leave and fertilize an egg, he will father most of the eggs she subsequently lays.

This double-barreled mating strategy increases a male's chance of fathering offspring and thereby enhancing his evolutionary fitness. As is to be expected, other species of insects have evolved similar innate behaviors programmed in the genes: some but not all wasps, bees, butterflies, and, as Chris Maier and I discovered,[5] several species of flower flies (family Syrphidae). Among the latter is *Mallota posticata*, which is large, hairy, yellow and black, and a very convincing mimic of a bumble bee. On sunny mornings in spring, females come to the blossoms of elderberry, wild rose, certain species of dogwood, and other shrubs to fill up on pollen, which they require as a source of protein to produce eggs. At this time of day, males make patrolling flights around these shrubs searching for females. If they spot one on a flower, they immediately pounce on and inseminate her. As noon approaches and the temperature goes up and the relative humidity goes down, most of these flies retreat to the nearby cool and moist forest. Here the females seek out wet rot cavities (tree holes) in tree trunks in which to lay their eggs. Males stake out a small territory around a tree hole and defend it against other males of their species. They dash out to intercept females in midair and copulate with them after the pair plummets to the ground.

Now back to the fruit flies. Both females and males, according to Ronald Prokopy and Daniel Papaj,[6] use both chemical and visual clues to locate and recognize the fruits in which females will place their eggs. Apple maggot adult flies can detect the odor of the fruits on a tree from a distance of 65 feet or more, and will follow this odor trail by flying upwind. (The attractive components of the odors of haw and apple fruits, now isolated and identified, have been formulated into artificial blends that are very attractive to these fruit flies.) Having arrived at a fruit-bearing tree, the flies must then locate a fruit. They do so mainly by means of visual clues. By hanging artificial models of fruits coated with a sticky substance in trees, Prokopy[7] and his coworkers found that the shape of the model fruit is the most important aspect of its gestalt. Similar in shape to a real fruit, spheres elicit the most attention from the flies. The size of the model has an amazing effect on the flies: the bigger the sphere the more flies gravitate to it and are trapped in the sticky Tanglefoot that coats it. Models the size of a softball, much larger than a fruit, trapped far more flies than smaller models that were the same size as a real fruit. The larger-than-life model is what ethologists, students of animal behavior, call a supernormal stimulus. (For example, Dutch ethologists found that herring gulls [which some refer to as *sea*gulls, although gulls are common on fresh-water lakes] are pushovers for larger-than-life artificial eggs. They nest on the ground and retrieve eggs that roll out of the nest. When presented with a choice of a normal egg and a grossly huge model egg, they retrieved the huge model. To a herring gull, the bigger the better.) Finally, the color of the model fruit is also significant. Early in the season, when fruits are yellow, the flies respond best to yellow models; later in the season, when the fruits are ripening, they prefer dark colors such as red. (Some people think that insects are color-blind, but they see colors very well. As a matter of fact, they can see ultraviolet, which is invisible to us.)

Once she has settled on a fruit, a female must decide whether or not to lay an egg in it. Is it a favorable habitat for the larva that will hatch from the egg? She penetrates the fruit with her ovipositor and makes her decision on the basis of chemical cues perceived by taste receptors on the ovipositor. Females will actually penetrate models made of soft wax, and will even lay an egg if the model has been flavored with sugar or, odd as it may seem, table salt.

This seemingly arcane and certainly esoteric knowledge of the apple maggot's egg-laying behavior seems, at least at first glance, to have no practical application. But that is not so. The aphorism noted earlier, "Know your enemy," is excellent advice whether the enemy is an invading army or a destructive insect. Sticky models that attract and trap fruit flies are hung on trees in apple orchards. They are red plastic spheres coated with Tanglefoot, a trapping adhesive. Just above each model is a vial that releases the volatile chemicals that attract flies to natural fruits from a distance. By trapping adult flies, these models give warning of the appearance of the first apple maggot flies of the season, an event whose timing varies from year to year. The number of flies caught is also a measure of the size of the population. This is useful knowledge to orchard owners. It tells them *when* an insecticide should be applied and whether it is *economically justified* because enough flies are present to constitute an economically significant threat to their crop.

In 1972 Ronald Prokopy[8] described a brief but very important behavior of female apple maggots. "Immediately following oviposition," he said, "apple maggot females were observed to circle around the fruit for about 30 seconds, dragging their fully extended ovipositors on the fruit surface behind them." Prokopy's field and laboratory experiments showed that they were marking the fruit with a pheromone—now known as the host-marking pheromone—that deters other females from inserting eggs into that fruit, thereby protecting their progeny from competition from other maggots. Since then, much more has been learned about this behavior. It is, as summarized by Francisco Díaz and coauthors,[9] characteristic of many but not all members of the fruit fly family. To varying degrees, related species of fruit flies recognize and heed each other's host-marking pheromones. This pheromone also stimulates flies to move elsewhere from heavily infested areas that have too many competing individuals, giving larvae even more protection from competition. (In some species, the host-marking pheromone induces males to remain in its vicinity, apparently signaling them that females are nearby.) The olive fruit fly, a serious pest of the olives that so often grace the tables of Mediterranean people, in 2002 struck the olive groves of southern Italy with a vengeance. This pest does not have a host-marking pheromone, but instead employs the juice of the olive fruit itself to ward off competing females. After she has pierced the fruit and inserted her egg, she uses her mouthparts to spread over the olive the juice that exudes from the egg puncture.

A species persists only as long as its members continue to produce offspring. The survival of a species is assured as long as each of the members acts to increase its own evolutionary fitness by striving to be survived by as many of its own offspring as possible—offspring that will pass the parents' genes on to future generations. As you read on you will find that one way to accomplish this is by producing very large numbers of offspring, which receive little or no parental care. In this way at least a few are likely to survive. An Atlantic salmon, for example, lays thousands of eggs in a shallow depression it scrapes in the bottom of a stream, but provides no care except to sweep sand and gravel over the eggs to hide them from predators. The aphids, which you will meet next, also use the strategy of producing very large numbers of descendants, but accomplish this in a way that is virtually unique among animals.

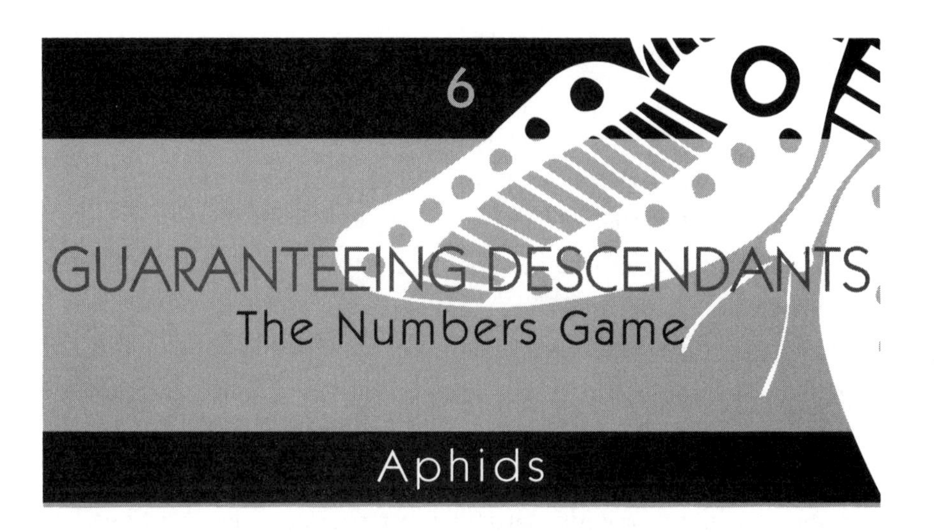

6 GUARANTEEING DESCENDANTS
The Numbers Game
Aphids

As she hangs upside down from a twig, a pert and petite black-capped chickadee—for the moment forsaking larger prey such as caterpillars—picks one aphid after another from the undersides of a spray of maple leaves. It will take many of these tiny sap-sucking insects, often called plant lice, to make a meal, but because they are exceedingly numerous, the bird compensates for their small size by eating a great many of them. No one knows how many, but two common yellow throats, warblers about the same size as a chickadee, have been seen to eat about 7,000 aphids in an hour. Our chickadee may be even more ravenous, because she requires extra nutrients to form the seven eggs that she will soon lay in a nest in a cavity that she and her mate excavated in a rotting birch stub.

The many species of woodland aphids are an ecological plus because they are an important food resource for chickadees and some other forest birds as well as for various insects. Despite the aphids' immense potential for population increase—more about this later—they seldom become destructively abundant in woodlands or other wild areas, because their food plants are difficult to find, scattered among other plants that they spurn, and because their natural enemies take a heavy toll of them.

A different species, the pea aphid (*Acyrthosiphon pisum*), not always so constrained, sometimes does serious damage to garden peas, especially to extensive commercial plantings with plants closely crowded together. The aphids pierce the stems, leaves, blossoms, and pods of the plants to suck their sap, often causing them to wilt; and they may also infect them with a viral disease. Because peas are planted early in the spring, the pea aphids' natural insect enemies—ladybird beetles, aphidlions, and larvae of certain hover flies—may not yet be numerous enough in some fields to keep the aphid population in check. Given this and the virtually endless supply of food, an aphid population may undergo a meteoric increase, spreading through the field like wildfire and all but wiping out the crop, and causing a disastrous loss of yield.

If a rapidly increasing population of pea aphids threatens a field, the only recourse is to spray the field with a contact insecticide, which kills when an insect touches it, before the aphids become so numerous as to be intolerably destructive. I was in charge of doing just exactly that during two long-ago summers when I was the entomologist for a canning company in Wisconsin. Pea aphids were a potential threat to the many fields of peas that nearby farmers grew on contract for the cannery. My assistants and I kept close track of these fields, frequently checking each one for aphids. We took ten or more randomly chosen samples scattered over each field by repeatedly placing a one-square-foot board between two rows, bending the adjacent plants over the board and shaking them vigorously to knock aphids down onto the board. If the average of all samples exceeded the critical number of 50 aphids per board, indicating a potentially economically significant population, the field was sprayed.

But why go to so much trouble; why not just spray all of the fields? The reason is that pea aphids—as is true of all crop pests—are not necessarily present in destructive numbers in all fields. For example, natural enemies may be more abundant in some fields than in others; and in some fields a fungus disease may decimate the aphid population. Routinely spraying crops that may or may not be threatened by insects is a waste of money and effort and, more importantly, needlessly pollutes the environment and accelerates the rate at which insects become resistant to insecticides.

The peas were sprayed from the air by a powerful Steerman biplane that laid down swaths of insecticide as it flew a series of straight passes over the field. Our usual practice was for a person at the far end of the field to wave a flag to mark the course of each pass, so that no unsprayed gaps were left between swaths. One pilot, who proved to be overly confident, assured me that he needed no guidance from a flag to spray a field without leaving gaps. He was a daring pilot who flew so low over the crop that he often returned with pea vines tangled in his landing gear. I thought he might just be able to do it. A few days after he sprayed, we sampled for aphids along a transect that crossed his swaths at right angles. As we sampled our way across the field, we found economically insignificant numbers of aphids in many areas, but in several places there were huge, seriously destructive populations, evidence that the pilot had left some unsprayed gaps and testimony to the aphid's astonishing potential for rapid and prodigious population increase.

How is it possible for aphids to produce such immense numbers of offspring in such amazingly short periods of time? In 1930 Robert Snodgrass of the US Bureau of Entomology wrote:

> "Plant lice!" Ugh, you say, "Who wants to read about those nasty things! All I want to know is how to get rid of them." Yes, but the very fact that those soft green bugs that cover your roses, your nasturtiums, your cabbages, and fruit trees at certain seasons reappear so persistently, after you think you have exterminated them, shows that they possess some hidden source of power; and the secrets of a resourceful enemy are at least worth knowing—besides, they may be interesting.[1]

That "hidden" source of power, according to Nancy Moran,[2] is their remarkable life cycle and their extraordinary means of reproduction. The seasonal cycle of many aphids, but by no means all of the 4,000 species, involves alternations between wingless and winged generations; between host plants, usually between woody and herbaceous species; and between parthenogenetically (without benefit of a male) and sexually reproducing generations.

The life history of the sap-sucking rosy apple aphid—which often damages the buds, leaves, and developing fruits of apple trees in spring—is complex, well understood, and especially intriguing. It survives the winter in the egg stage, tucked into crevices in the bark of apple twigs or branches. The shiny, ovate, dark green eggs hatch in spring when the buds burst open. Every egg produces a wingless female that is parthenogenetic and gives live birth to her offspring. (No males will be born until the last generation of the season.) She is known as the "fundatrix," or stem mother, because she founds a dynasty of fifteen or more generations that span the warm months of the year. After maturing, the stem mother bears several dozen wingless daughters, all exact copies of herself—clones. Like their mother, they are wingless, parthenogenetic, and give live birth to a few dozen females that are exact copies of themselves—as will be their daughters, granddaughters, and all succeeding generations except the last. Later in the year, in early summer, many females that will grow wings are born. After they feed and mature, most will fly to their alternate host, a common lawn weed, the narrow-leaved plantain. On this plant there will be several generations of wingless, parthenogenetic, live-bearing females, all clones of their mothers and, of course, of the long-gone stem mother. In autumn winged females reappear, migrate to an apple tree, and give live birth to the one and only generation of sexually reproducing females, which are wingless. Winged males are born on the plantains, and as explained by L. J. Pickett and coauthors,[3] fly to an apple tree, attracted by a sex pheromone emitted by the females, and mate with a female who will start a new cycle by laying a few eggs that will hatch in spring—usually less than six—in crevices and cracks in the bark of the apple tree.

Chickadees and aphids have two distinctly different reproductive strategies for enhancing their evolutionary fitness. Both increase the probability that some of their offspring—the more the better—survive to pass their parents' genes on to future generations. The aphids, like many other insects and some other animals, produce many progeny—so many that they can give them little or no parental care. Their genetically programmed strategy is to gamble that a few of the many will survive. In contrast, chickadees, like other birds and mammals and even a few insects (one of which you will soon meet), produce only a few offspring but enhance their chance for survival by giving them a large measure of parental care. While aphids give their young no care, chickadees faithfully incubate their eggs, brood and feed their nestlings, and even continue to feed them for two to four weeks after they leave the nest.

Aphids and birds are near opposite ends of a spectrum of reproductive strategies. At one extreme are organisms such as the codfish that lay and abandon as many as six million eggs, which contain the bare minimum of life-sustaining yolk (known as the r-strategy among ecologists). At the other extreme are animals such as elephants and people that bear very few young but nurture them during a long period of gestation and then protect and feed them for years (known as the K-strategy among ecologists). Most animals are somewhere between the two extremes, as are chickadees and other birds, and, closer to the other extreme, insects such as the tobacco hornworm moth that invests her eggs with a relatively generous portion of yolk, and does not abandon them until after carefully gluing them one at a time to leaves of tomato, tobacco, or other plants of the nightshade family—the only plants her host-specific caterpillars are adapted to eat.

Most species that follow the strategy of prolific breeding reproduce sexually generation after generation. A female inseminated by a male produces a horde of offspring, each one containing genes from both the father and the mother. The pea aphid, rosy apple aphid, and many other aphids achieve the same end indirectly. The parthenogenetic stem mother gives rise to a cascading dynasty of generations, parthenogenetically reproducing clones of herself that have only her genes—none from a male. With some luck, her clonal descendants will become more numerous with succeeding generations. The members of the last parthenogenetic generation, genetically identical to the founding stem mother, give live birth to a generation of sexually reproducing males and females which are, in the aggregate, the equivalent of the great horde of eggs laid by a sexually reproducing codfish, and that will themselves lay just a few eggs each.

Multiplying herself through a parthenogenetically reproducing dynasty of clones rather than by reproducing sexually and by herself laying a large number of eggs can be a very successful strategy for a stem mother. Assume that a stem mother bears about 50 young (conservatively low for most species) before she dies, and that all of them and their daughters, granddaughters, great-granddaughters, and so forth survive. After six generations, less than half the usual number for aphids, the stem mother would have about 15 billion descendants—all clones, exact genetic copies of herself. This is, of course, impossible. However, we can more reasonably assume that only two daughters of each generation survive. Even so, after 15 generations, the stem mother would be survived by over 32,000 descendants that are all clones of herself, and that will, in the last parthenogenetic generation, give birth to the final generation—sexually reproducing males and females. The latter, the proxies for the original stem mother, will lay the eggs, which are the only representatives of the species present in winter. Each female will lay only a handful of eggs, but there will be so many of these sexually reproducing descendants of the stem mother that the total is likely to be in the thousands. But there is not likely to be a population explosion. Few of the eggs will survive the winter because of the toll taken by parasites and predators such as bark-creeping nuthatches, brown creepers, and other birds.

It is not unusual for aphids in the temperate zone to have as many as 20 generations during the warm season, and in the warmer climate of southern Florida, the citrus aphid may have as many as 47 in a year. In either case, a new generation is born every seven to eight days. This greatly exceeds the growth rate of most insects. How can aphids accomplish this amazing feat? There are several ways.

First, parthenogenetic reproduction makes possible a "telescoping" of generations. That is, an embryo developing within her mother already has developing embryos, the grandchildren of the mother, in her own body. This unusual phenomenon was succinctly explained by A. F. G. Dixon:

> Sexual reproduction precludes the telescoping of generations because an animal that must mate cannot begin to mature its embryos before it is born. On the other hand, the embryos of a parthenogenetically reproducing aphid can have embryos developing within them. This shortens the period between the appearance of adults in successive generations and that between the adult moult and the onset of reproduction, both of which also contribute to the prodigious rate of increase achieved by aphids.[4]

Furthermore, once born, the aphids develop to reproductive maturity very rapidly, according to data compiled by Dixon, at temperatures between 68° and 77° F, in as little as 6 days and even at low temperatures of between 50° and 59° F in 17 days. Finally, growth is enhanced because the overall nutritional quality of the aphids' diet is improved by the seasonal alternation between a woody and an herbaceous host plant. The leaves of woody plants are highly nutritious in spring but far less so in midsummer. But by midsummer the aphids have migrated to an herbaceous host, which is likely to be considerably more nutritious.

Although the aphids we have met so far—as well as many other species—have a sexually reproducing generation, quite a few others, especially tropical species, reproduce only parthenogenetically. Why reproduce sexually when parthenogenesis is so much more rapid and efficient and does not require that a male and a female find each other—which does not always happen? Put briefly, the answer is that sexually reproducing species are more likely to adapt to changes in their environment because sexual reproduction yields progeny that vary in structure, physiology, and behavior. Variation is the grist for the mill of natural selection, because some progeny may be better able than others to cope with the demands of the environment. On the other hand, parthenogenetically produced individuals are generally exact copies of their mothers; they look and act alike, showing little or no variation. They are specialists adapted to cope with their environment as it is, and are prone to extinction if the environment changes. As Ursula Goodenough, a cell biologist, put it in her wonderful *The Sacred Depths of Nature*, "specialization [as a result of parthenogenesis] can definitely be a good idea over the short haul, when a particular facet of an [ecological] niche can be exploited by a particular kind of creature. But it is vulnerable to the fact that most niches keep changing."[5]

It is obvious to all that almost never are two members of a sexually reproducing species exactly alike. Witness the fact that we humans can tell apart virtually all other humans with the exception of identical twins, which are of course clones of each other. We have so much potential for variation that we have adapted to virtually all parts of the world, from the Arctic to the tropics. Where does all of this variation come from? It is tucked away in our genetic makeup as it is in that of all other organisms. A great many of the thousands of a species' genes are cast in somewhat different forms in different individuals. These alternate forms, known as "alleles" to geneticists, cause slight or even major changes in the characteristics they affect.

While the offspring of a parthenogenetic female have only one set of genes, those obtained from their mother, the offspring of a sexual union have two sets of genes, one from the father and one from the mother. These two sets correspond functionally gene for gene, but many of these genes will exist as different alleles in different individuals. The genetic makeup of an individual is thus likely to include a unique combination of alleles. For example, an allele affecting the sense of smell may coexist with an allele affecting the length of the antennae; a combination that may be rare or even unique—and may under some circumstances be very useful.

While aphids and most other insects play the numbers game to a greater or lesser extent, a few insects, as noted, use the opposite approach—bearing only a few young but giving each one a good measure of parental care, as do birds and mammals. Among these insects are several species of flies, including the dreaded tsetses, the subject of the next chapter, which transmit the debilitating and usually fatal sleeping sickness of Africa. Female tsetses, to the astonishment of most people, protect their larvae by retaining them in their body for a long period of gestation, until they are full grown, may weigh more than the mother, and are ready to molt to the pupal stage. While in the mother's "uterus," the larva subsists on milk that is secreted into the uterus. Read on to learn more about the amazing similarities between reproduction by the tsetse and humans and other mammals.

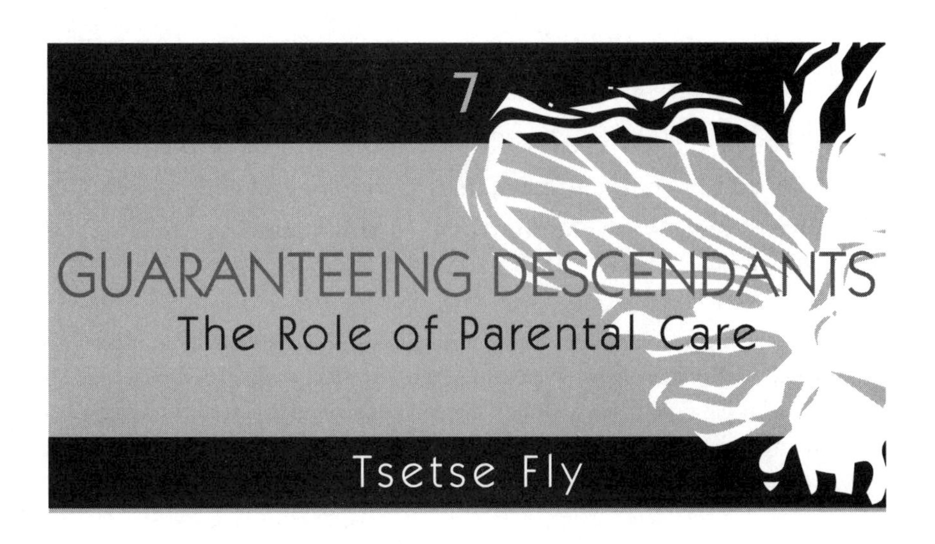

7

GUARANTEEING DESCENDANTS

The Role of Parental Care

Tsetse Fly

The mother labors to give birth to her baby. Periodic contractions course through her body, infrequently at first, but more and more often as the moment of birth nears. Finally, a more rapid series of contractions signals the imminent emergence of the baby from the uterus. Although this sounds like a typical human birth, it is not. In this particular case the baby is an insect, the full-grown larva of a tsetse fly, and this account of its birth is based on the observations and experiments of David Denlinger and Jan Ždárek.[1]

Female tsetses (genus *Glossina*) and a few other flies have an analogue—not a homologue—of a mammalian uterus, a large swollen part of the egg canal. Instead of the placenta that nourishes fetal mammals, a pair of milk-secreting glands that empty into the "uterus" feed the developing larva. Tsetse milk is white and chemically similar to human or cow's milk. Like mammalian milk, it is a nutritionally complete food that provides all of the nourishment required by a developing offspring.

Although the term "tsetse fly" is frequently used in the scientific literature and in common usage, it is a tautology; that is, appending this insect's name with the word "fly" is redundant. As explained by Patrick Buxton:

> The word tsetse comes from the Sechuana language, spoken by some of the people of Bechuanaland [Botswana], and signifies a "fly destructive of cattle." It has been suggested that "tsetse" is onomotopoeic, the sound suggesting a sharp buzz. . . . The word was first used in English in 1849, and very soon became general in traveler's books. . . .[2]

Tsetses, all of them bloodsuckers, are indeed destructive of the cattle that are the mainstay of many native African cultures, but the flies are also inimical to humans in more direct ways. "Of all the biting flies," explains Robert Snodgrass,

> there is none to compare with the tsetse fly of Africa. . . . Not only is this brownish fly, not much bigger than a house fly, an intolerable nuisance to men and animals because of the severity of its bite, but it is a deadly menace by reason of its being the carrier of the parasite of African sleeping sickness of man, and that of the related disease called *nagana* in horses and cattle.[3]

The several species of blood parasites that cause these diseases, single-celled creatures, protozoans known as trypanosomes, are transmitted from wild animals such as buffalo, zebra, wildebeest, and various antelopes to humans or cattle mainly by the bite of either a male or female of only 4 of the 21 species of tsetses.

According to Maurice James and Robert Harwood, *nagana*

> has denied man the use of domestic animals except poultry over an enormous area [south of the Sahara] estimated to be as great as one-fourth of the [entire] African continent. The result is not only a loss of the much-needed protein supply, so essential to the future development of an already protein-starved continent; there is also the influence on the [customs] of African peoples of certain tribes [e.g., the Zulu] to whom possession of cattle is an important status symbol.[4]

The bite of a tsetse not only transmits trypanosomes that cause sleeping sickness in humans—as you will soon see—but its bite is so painful and annoying that, in infested areas, people carry whisks that they keep in almost constant motion to brush the flies away from their bodies. Taking a lesson from nature, they usually make whisks from the hairy tail with which wildebeests vigorously lash themselves to chase off tsetses and other biting flies.

A tsetse that gets by the whisk and bites may inoculate its victim with one of the trypanosomes that cause sleeping sickness, a profoundly debilitating disease that ultimately results in death. After an incubation period of 10 to 20 days, the trypanosomes, which are then in the blood, cause bouts of fever, weakness, and lassitude. After some months, the trypanosomes move into the cerebrospinal fluid. This final stage of the disease is characterized by nervousness and drowsiness that eventually give way to extreme sluggishness. The sufferer eventually becomes comatose, does not eat, wastes away rapidly, and finally dies. As you will see, the best "cure" is prevention by means of controlling tsetse populations.

Many insects locate mates by means of one of the three senses that can perceive signals from a distance: vision detects the flashing light of a firefly; female crickets hear the song of a male; and male bagworms use their sense of smell to orient to the female's sex attractant pheromone. Other insects, tsetses among them, meet their mates in places where members of the opposite sex are likely to be in the course of their daily lives. People meet their potential mates in many places: singles bars, libraries, laundromats, or churches, to mention just a few. But insects lead far simpler lives, and males—almost always the sex that searches for a mate—are likely to find females in only two trysting places: where the females eat or where they lay their eggs. Certain hover flies, as you will read, ambush females as they approach the tree cavities in which they lay their eggs. Some solitary bees wait for females to come to the particular blossoms from which they sip nectar.

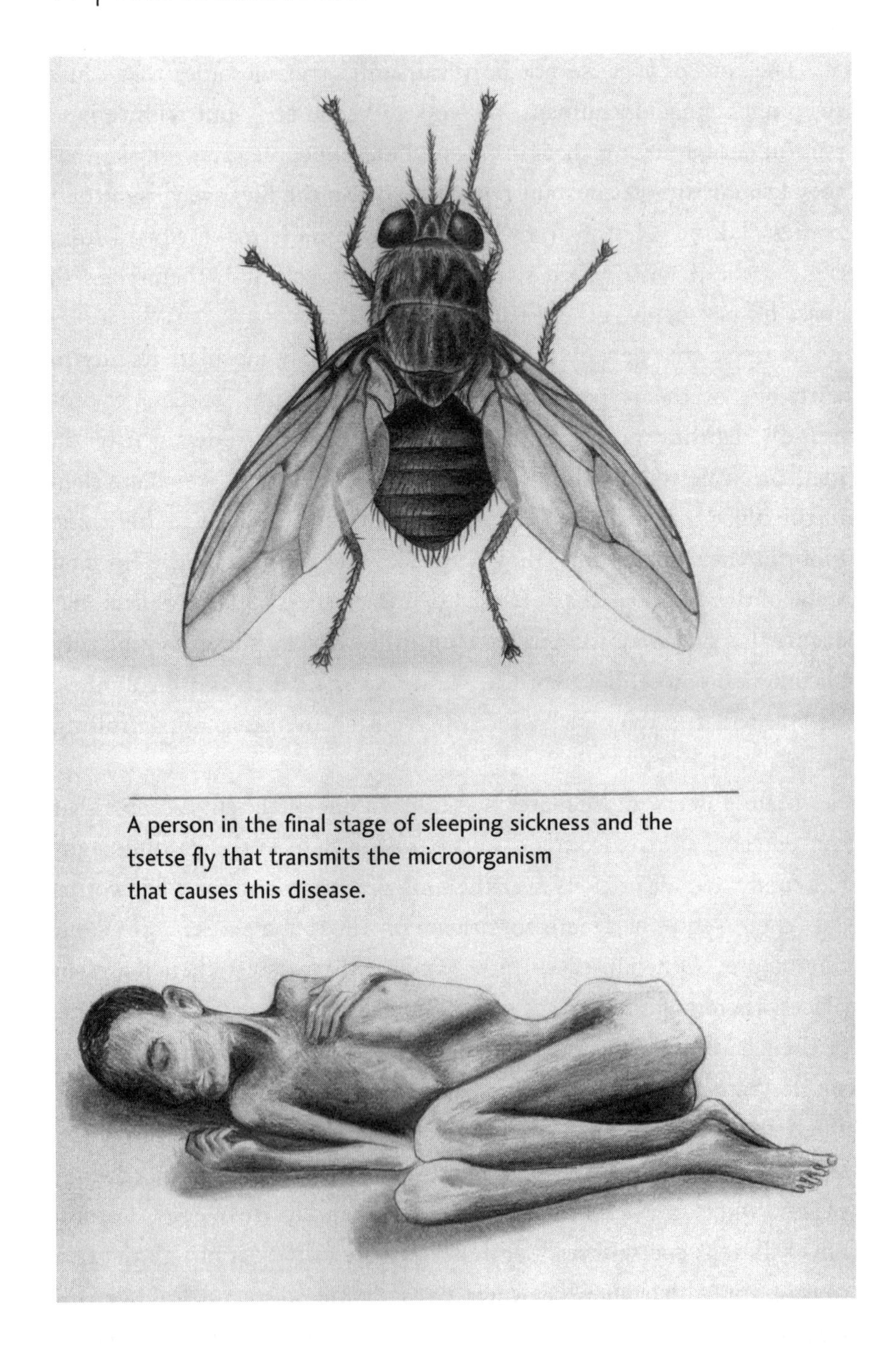

A person in the final stage of sleeping sickness and the tsetse fly that transmits the microorganism that causes this disease.

Male tsetses follow or perch on large grazing mammals such as wildebeests and zebras and try to mate with females that arrive to take a blood meal. In an experiment done years ago, male tsetses responded similarly to a vehicle that was draped with a gray blanket and driven slowly across the savannah. They followed the vehicle as it moved and alighted on the blanket when the vehicle stopped. It appears that male tsetses rely on dimly perceived visual cues to identify the large animals that are their trysting sites. According to a news report by Josh Gewolb in a recent issue of *Science*,[5] entomologists in Zimbabwe are using a similar tactic to control tsetses. Beginning in 1984, they deployed "fake cows" consisting of a hanging rectangle of black and blue cloth impregnated with a deadly insecticide. Over 60,000 "fake cows" now dot cattle ranches in Zimbabwe, and the rate of infection in real cows has plummeted from 10,000 per year to only 50 per year. The infection rate is nearly nothing. The problem has been solved.

Gestation begins when an egg, only 0.06 inch long, descends from the ovary into the female's uterus, where it is fertilized by a sperm from a previous copulation. (The male had ejaculated semen into her uterus through his penis and she had stored it for later use in two sacs, the spermathecae.) The egg soon hatches and the larva feeds as it clings to the orifice of the milk glands. During gestation the mother must take three blood meals in order to keep up the flow of milk. After 9 or 10 days, the larva, having molted twice, is full grown and ready to be born. When she is ready to give birth, the mother seeks out a suitable site, generally but not always, a place where the larva can burrow into soft, moist soil. The birth usually occurs in late afternoon, takes from one to two hours, and as Denlinger and Ždárek[6] observe: "The fact that the fully grown larva within the uterus of the mother can actually weigh more than the mother suggests that the process of giving birth is not a trivial event." After parturition the mother makes a distinct buzzing sound that Denlinger jokingly calls the "post partum blues." But it might really be a "sigh of relief."

Immediately after its birth, the larva burrows into the soil if it is loose enough, and within minutes forms a puparium. If the soil is hard and impenetrable, the larva wanders about. If it finds a suitable site it burrows into the soil or under debris, otherwise, it forms the puparium on the surface, where it is exposed and likely to be found by a predator. After about three weeks, the adult fly sheds its pupal skin, pushes its way out of the puparium, and forces its way up through the soil. Buxton notes that "when it first reaches the surface, the wings are not yet expanded and the fly cannot take to the air; but it runs actively over the surface of the ground."[7] After about an hour the wings have expanded and the tsetse is capable of flying. Nevertheless, it remains inactive for up to a day or two before it flies off to take its first blood meal.

Among the insects, tsetses are K-strategists par excellence, that is, they seek to perpetuate their genetic line by bearing few offspring but try to ensure their survival by investing in each one a great deal of parental care, in their case, a long period of gestation. Data summarized by Buxton[8] indicate that female *Glossina morsitans*, probably the most important vector of *nagana* and sleeping sickness, survived in the laboratory an average of 72 days. By marking large numbers of them, releasing them in the field, and then recapturing as many as possible, it was found that in nature, where they are exposed to predators and other dangers, the tsetse's life span is considerably shorter, a maximum of 61 days for a female. Males survived only half as long, presumably because they are more exposed to predators since they are more active.

At an average laboratory temperature of 86° F, the gestation period is almost eight days. (In the hottest month in Nigeria the average outdoor temperature is 89° F; in a rainy month it is 77° F.) Assuming highly favorable conditions: a mean temperature of 86° F, a 61-day life span, a gestation period of 8 days, and an interval of 1 day between pregnancies, a female could give birth to a maximum of no more than six young during her lifetime. This is even less than the average number of eggs laid by the black-capped chickadees discussed earlier, about 17—an average of 6.7 eggs per year during an average life span of 2.5 years, according to Susan Smith.[9]

A few other insects reproduce like tsetses. Among them is a parasite of sheep commonly known by the misnomer sheep "tick" that is found on sheep and goats throughout the world. Although it is vaguely ticklike in appearance and behavior, it is actually a bloodsucking fly that is radically adapted to spend its entire life on the body of a sheep. More aptly called sheep keds (*Melophagus ovinus*), these flies have lost their wings, are flattened like a tick, and have their legs spread to the side better to grip the host. (According to Robert L. Metcalf and Robert A. Metcalf,[10] the sheep ked "causes sheep to rub, bite, and scratch at the wool, thus spoiling the fleece. . . .") If sheep keds did not retain their larvae in the "uterus" until full grown, they could not live as permanent parasites on sheep, or any other animal. A wingless adult sheep ked would find it exceedingly difficult to find a sheep and to climb up onto its body if it spent its larval and pupal stages somewhere other than on its host—very likely in carrion or dung, as do most fly larvae and as did the ancestor of the sheep ked. Even if it were capable of boarding a sheep, the roaming flock would surely have moved on by the time the free-living larva had matured to the adult stage.

A sheep ked's "uterus" and milk glands are very similar to a tsetse's and as Beth Lenoble and David Denlinger[11] discovered, the milk they secrete is also chemically very similar. The problem of spending an entire lifetime of up to five or six months on the sheep is solved when the mother ked fastens her newly born and full-grown larva to the sheep's fleece with a gelatinous glue that she secretes. The larva quickly pupariates, molts to the pupal stage within the puparium, and emerges—on the sheep—as an adult about three weeks later.

An animal must survive to adulthood if it is to bear progeny that will survive it—whether it gives parental care or plays the numbers game. In the temperate zone, every animal is faced with surviving the cold of winter. Like many birds, a few insects, such as the monarch butterfly, escape the cold by migrating south to a warmer climate. Most survive by going into a period of "winter sleep" (diapause) that, as noted earlier, is similar to but not the same as the hibernation of mammals such as bears and woodchucks. Their metabolic rate is greatly lowered and their bodies are resistant to freezing. It is the same with the eggs, larvae, pupae, or adults—which one depending upon the species—of virtually all temperate zone insects. As you read on you will find out how the bagworm, a common pest of ornamental trees and shrubs, survives the winter.

8

SURVIVING WINTER AS A SLEEPING EGG

Evergreen Bagworm

Strolling in your yard on a sunny day in December, you stop to admire the row of flourishing arbor vitae (northern white cedar) trees near the edge of your property. You notice something you had never seen before, about three dozen brown, spindle-shaped bags hanging from branches. Each sturdy, silken bag is about two inches long and festooned with plant fragments. Last summer these bags were the cocoonlike cases, mobile homes, in which lived voracious caterpillars that feasted on the leaves of your arbor vitaes. Some of them, those that had housed a caterpillar that became a male moth, are now empty. But bags that had been home to a female contain a molted pupal skin packed with, on average, over 1,000, but occasionally over 2,000, tiny, white eggs, all now deep in their winter sleep. Assuming that about half of the bags contained females, when the eggs hatch in spring your arbor vitaes will be threatened by as many as 18,000 hungry caterpillars, enough to do unsightly damage or even defoliate and kill the trees.

How can you save your trees from certain destruction by such a hungry horde? Do not hire a professional exterminator! This is one of the many insect problems that can be controlled without using insecticides. With but little effort, you can forestall the impending threat to your trees by picking these few bags by hand and discarding them in the trash can. If you do this every fall or winter, there will be few or even no bagworms on your trees. Of course, if the trees are too tall you may have to climb or spray an insecticide in spring when the eggs hatch.

This insect, the evergreen bagworm (*Thyridopteryx ephemeraeformis*), is a pest in much of the eastern United States; its range extends in the north from Connecticut to Nebraska and in the south from Georgia to Louisiana. When the eggs hatch depends upon the latitude—earlier in the year in the south than in the north. Robert Morden, my former graduate student, and I[1] found that the date of hatching is fine-tuned by the weather. In Urbana, Illinois, the period of hatching extended from May 29 to June 9 in a cool spring and in a warm spring from May 21 to June 1.

A newly hatched bagworm caterpillar, a tiny acrobat and sometimes an aeronaut, is well worth watching. After escaping from its maternal bag, it moves about actively with its abdomen stuck straight up, almost as if it were standing on its head. When it finds a suitable place on a leaf, it immediately begins to construct the bag in which it will live for the rest of its larval life—much as a hermit crab inhabits a discarded sea shell. It begins, as Tohko Kaufmann found in a study of a related bagworm's behavior,[2] by spinning a girdle of silk supported by two silken "guy wires" that slant to the leaf from opposite sides of the girdle. The caterpillar pokes its head through the girdle and adds silk and tiny bits of plant material to its front margin as it slowly crawls forward. It soon completes a tiny sleeve less than 0.2 inch long and with an opening on each end. Because the sleeve is then not long enough to cover all of the caterpillar, the head protrudes from the front end and the tip of the abdomen from the other end. The little sleeve is eventually completed and becomes a proper bag that is roomy enough to accommodate the whole caterpillar. It has a front opening through which the legs and head protrude when the caterpillar is crawling or eating and a rear opening through which fecal pellets are discarded. The bags are similar in appearance but can vary in size depending upon the size of its occupant, which in turn depends upon its age and the nutritional quality of the leaves of the species of plant on which it is feeding.

Some recently hatched bagworms may disperse from their home site by "ballooning" on a strand of silk. In their research on this insect's means of dispersal, David Cox and Daniel Potter[3] discovered that after constructing the initial sleeve—occasionally before doing so—some of the tiny caterpillars lower themselves on a thin, single strand of silk that may be less than an inch long or as much as two feet long. If the strand is torn from its mooring by a breeze, the caterpillar holding the other end will be wafted away like a kite, only a few inches if the strand is short, or as much as 250 feet if the strand is long! Ballooning has its risks. If the caterpillar falls to the ground or lands on the side of a building, it will starve to death unless it can make its way to a food plant. If it lands on a woody plant, a tree, or a shrub, it will probably be in luck because, unlike many insects that will accept only a few species of plants, the bagworm feeds on most woody plants, both evergreen and deciduous ones, which lose their leaves in winter, although it has a strong predilection for arbor vitae, juniper (red cedar), and other conifers.

Dispersal is advantageous for bagworms and virtually any other organism, because it prevents crowding and competition for food with parents and siblings, offers an escape from natural enemies, and makes possible the colonization of new areas. In most moths and butterflies, as well as almost all other insects, dispersal is accomplished by the mother as she flies from place to place to scatter her eggs. But because their mothers are wingless, bagworms can disperse only when they are in the mobile larval stage.

As explained by Leonard Haseman[4] of the Missouri Agricultural Experiment Station, as bagworm caterpillars gradually grow from a length of less than 0.2 inch to about 1.2 inches, they must continually enlarge their bags by adding more silk and more plant fragments to the front edge. As they increase in size, the caterpillars molt their skins six or seven times. In preparation for a molt, the caterpillar fastens the front end of its bag to a twig by a sturdy, encircling strap of silk and then shuts the front opening with silk. As best we know, female caterpillars attain their maximum size after the first six molts and males after the first five. In either sex the next molt is to the pupal stage. In preparation for this last larval molt, which occurs in the fall, the caterpillar again fastens its bag to a twig and closes the front opening. It strengthens the bag by adding more silk to its inner wall and then turns around so that its head points downward, facing the rear opening through which the fecal pellets had been discarded. In two or three weeks, the pupa undergoes a final molt, the molt to the adult stage, the moth. Eggs laid by the resident female will spend the winter in this bag.

In their pupal and adult forms, the sexes of the bagworm are radically different in both appearance and behavior. However, "during larval life," writes Haseman, "there is little external difference between the two sexes except for size. On preparing to pupate, the female bag is about twice the size of that of the male and the female caterpillar is much larger than the male."[5] The pupae are even more different. Males have externally visible legs, wings, and antennae, all fused to the body of the pupa, thereby giving it the appearance of an insect carved in bas relief on a small nut. The female pupa, on the other hand, although segmented, as is the male, is otherwise just an oblong brown sac lacking all appendages.

Female caterpillars on deciduous trees have a problem that those on evergreens do not have. In autumn deciduous trees drop their leaves but evergreens do not. If a female fastens her bag to a woody twig of a deciduous tree, the bag and the eggs in it will remain in the relative safety of the tree until the following spring. But if she fastens it to the stem of a leaf, it will fall to the ground with the leaf, exposing her eggs to hungry mice and probably to fungal infections throughout the winter. On the other hand, it does not matter if the empty, discarded bag of a male falls from the tree. Peter Lagoy and Edward Barrows[6] showed that, as expected, females are much more discriminating than males in choosing safe attachment supports. Over 97 percent of 395 females on the deciduous black locust anchored their bags to woody twigs about 3 millimeters or more in diameter, while about 77 percent of 158 males attached their bags to leaf stems, which, unlike twigs, will fall from the tree in autumn and are only about 1.5 millimeters in diameter. Lagoy and Barrows concluded that females identify a suitable attachment site by its thickness, which stands them in good stead, because twigs are almost always thicker than leaf stems.

Jim Sternburg and I[7] found that promethea caterpillars, innocuous insects beloved by naturalists, are as discriminating as female bagworms in fastening their cocoons to a twig. Promethea caterpillars are naked, but both sexes survive the winter as a pupa in a silken cocoon dangling from a woody twig of their host tree, which may be, among a few others, wild black cherry, sassafras, or ash. In autumn the caterpillar spins a cocoon that must be firmly attached lest it fall to the ground and be ravaged by a hungry mouse. The cocoon is wrapped in a leaf, and both leaf and cocoon are attached to the adjoining twig by a strong strap of silk. The strap sheaths the nonwoody stem of the leaf, extends to its base and onto a variable length of the woody twig, to which it is firmly attached. On thick cherry twigs the silken attachment to the twig is somewhat more than a half-inch long, but on thin cherry twigs the average length of this attachment is over three inches, and it may be even longer and extend beyond the next fork of the twig. The seemingly unnecessarily long attachment to a thin twig that will not fall from the tree probably reflects a behavior that prevents cocoons from falling from trees, such as ashes, which do not have a simple leaf, but rather a compound leaf, several leaflets attached to a long nonwoody stem that falls in autumn. On ash the caterpillar wraps its cocoon in a leaflet, but prevents itself from falling from the tree with the compound leaf in autumn by extending the anchoring strap up the long and relatively thin leaf stem to the much thicker adjoining woody twig.

Robert Morden and I[8] observed that in Urbana, Illinois, male bagworms emerge from the pupa in late September and early October, between noon and six in the afternoon. When an adult male is about to emerge, the pupa squirms about three-quarters of the way out of the bag and then the pupal skin splits along the middle of the top of the head and thorax. Still wet and with shrunken wings, the male moth forces its way through the narrow split and hangs by its legs from the now empty pupal skin. Soon he becomes dry and, by the force of blood pressure, the wings expand to their full size, a spread of as much as 1.5 inches. The moth is covered with a growth of almost black scales, but those on the wings soon fall off, leaving the membranous wings transparent like those of a fly or bee. The males are masterful and rapid fliers but usually survive no more than a day because they cannot feed by sipping nectar from flowers, as do many other moths, because their mouthparts are vestigial.

"No marked change," explained Haseman,

> takes place in the . . . pupal case [skin] of the female on maturing. An inconspicuous slit appears at the lower or head end of the pupal case, but the mature female insect remains inside the case after maturing. If one cuts open one of the larger bags . . . and then removes the mature female insect from the brown pupal case, he will find that she is simply a soft, slug-like object without any legs, wings, or other appendages . . . she is merely a fleshy bag filled with eggs.[9]

Her thorax and abdomen are covered with hairs—rather sparingly except for a dense girdle of yellowish brown hairs at the end of the abdomen. The female does not leave the bag until after she has mated and laid her eggs.

In the afternoon, when a female is ready to mate, she protrudes her head and thorax through the slit at the head end of the pupal case. In a fascinating description of the bagworm's mating behavior, B. A. Leonhardt and coworkers[10] wrote that she then expels a puff of golden thoracic hairs from her pupal case into the front of the bag. The hairs release a volatile sex attractant pheromone that wafts downwind and guides a searching male to her. He lands on her bag and then the pair copulates in an amazing way that may be unique in the animal kingdom except for relatives of the evergreen bagworm. Mating is not easy for bagworms. Haseman describes the act:

> The insect has a very peculiar method of mating. Since the female remains inside the pupal case, which is enclosed by the tough silken bag, it becomes necessary for the male to reach her inside this double barrier. There is an opening in the lower end of the bag and also a slit in the head end of the pupal case which is next to the opening. So the male moth, being able to extend its abdomen to fully three times its normal length, thrusts the end of it through the mouth of the silken bag and into the narrow slit in the pupal case and gradually forces it up between the body of the female and her enveloping pupal case until the posterior end of her body is reached and impregnation takes place.[11]

Haseman continues with a clear and concise description of egg laying:

> Soon after impregnation the female begins to deposit her supply of eggs in the upper end of the pupal case. There is no system to the arrangement of the eggs though they are closely packed together. They fill all of the abdominal portion of the case with the exception of the first large segment in front. . . . This and the thoracic cavity are filled with the soft scale-like pubescence from the posterior end of the female's body. As the eggs are being deposited, a quantity of this pubescence is worked in between them so as to isolate one from the other, though the majority of it is used as a plug for the front end of the case. From the nature of this pubescence and the way it is tamped into the front end of the pupal case, it offers an almost impenetrable barrier to any small enemy that might wish to reach the eggs.[12]

The mated female completely fills the pupal skin, and as she lays her eggs, she gradually shrivels up, so that by the time she is finished laying and she has tamped in the plug of scales, there is very little left of her. With a last effort before dying, she pushes her way out through the slit at the front of the pupal skin and usually through the opening in the bag and falls to the ground. After she leaves, the slit in the pupal skin closes and the eggs are safely sealed in the bag, which is waterproof and almost impenetrable and hangs pendant from branch or twig throughout the winter.

How do bagworm eggs survive the long and often bitter cold of winter? The bag may help just a bit, but don't think of it as a blanket that keeps in heat. Unlike you in your cozy winter bed, bagworm eggs produce no heat for their silken blanket to retain. Like most insects, bagworm eggs are in diapause. Diapause can occur in any life stage—egg, larva, pupa, or adult—but in a given species almost always in the same stage. In diapausing insects, physical development to the next life stage stops: eggs do not hatch until diapause has been terminated; larvae do not metamorphose to become pupae, pupae do not become adults; and adult females do not lay eggs. Insects in winter diapause usually have physiological mechanisms that allow their bodies to supercool. That is, their body temperature can fall below freezing without the formation of ice crystals that would irreparably damage the cells in the body and thus cause them to die. Furthermore, the metabolic rate of diapausing insects in any stage is far lower than that of nondiapausing insects, usually only one-tenth or less of the usual rate. An insect whose metabolic rate is down to one-tenth of its normal rate reduces its consumption of energy by 90 percent, and can thus survive on its stored energy ten times as long as a nondiapausing insect.

What, if anything, triggers an insect to go into diapause? A very few require no trigger; they automatically go into diapause at the appropriate stage of development, which seems to be the case with the evergreen bagworm. Entomologists call this an obligatory diapause.

Most insects have a facultative diapause. That is, unless they are triggered by some environmental factor, they continue to develop and without interruption give rise to the next generation. For species that grow rapidly and can have two or more generations in one summer, this is an advantage. In the laboratory Bob Morden and I[13] raised bagworms from egg to adult under various combinations of constant temperatures and artificial day lengths, but no combination prevented females from laying diapausing eggs. Therefore, we think that bagworms probably have an obligatory diapause, which makes sense, because they grow so slowly that they can have only one generation per summer. But, as Morden and I cautioned, "It is not easy to prove the occurrence of obligatory diapause, since it is always possible that some untried factor will prevent its occurrence. For example, although we found that a constant [day length] does not prevent diapause, it may be that a constantly increasing [day length] might do so."[14]

Like bagworms, domestic silkworms diapause in the egg stage, but their diapause is facultative. It must be triggered by some environmental cue that tells them that winter is approaching. The only completely reliable cue that denotes the seasons is the length of the day. Temperatures are not reliable because they are so variable. In the North Temperate Zone, December 21, the first day of winter is the shortest day of the year. On that day in central Illinois there are fewer than nine and a half hours of daylight plus a total of about half an hour of twilight at dawn and dusk. The first day of summer, June 21, is the longest day of the year, in central Illinois a little more than 15 hours of daylight plus a half hour of twilight. From June to December the days gradually become shorter, auguring the approach of winter. From December until June the days gradually become longer, foretelling the arrival of spring. Some insects are induced to enter a summer diapause, often to escape extreme heat or dryness, by the long days of late spring and early summer. But most insects in temperate areas have only a winter diapause, which they enter in response to the relatively short days of late summer and early fall.

Because undeveloped eggs have no eyes or any other way of sensing light, egg diapause must be initiated by the mother. In the domestic silkworm, which spins the silk in our shirts and blouses, the day length to which the mother was exposed when she was an embryo—embryos do have eyes—determines whether or not her eggs will enter diapause. While most insects go into winter diapause in response to short days and are prevented from doing so by long days, silkworms, "planning ahead" for offspring they will not produce until they become adults weeks later, respond to day length in the opposite way. The first generation of the year is in the embryonic stage in early spring when the days are still short. Nevertheless, they are predestined to lay nondiapausing eggs weeks later when they have grown to the adult stage. By then it will be summer, and their offspring will prosper during the long, warm days. The long days predestine their offspring to lay diapausing eggs when they become adults in late summer, eggs that will not hatch until the following spring. If they were to hatch so late in the season, cold weather would kill the caterpillars before they could complete their growth.

When the warm days of spring arrive, insects of the temperate zones resume development and reproduction. But different species do so at different times—some very early in spring and others much later—in order to coincide with the availability of their various resources, perhaps a specific food plant for an herbivore or a favorite insect host for a parasite. How do they know that spring has arrived or will shortly arrive and that it is safe to resume normal life? The first warm days of spring do foretell the approach of the warm season, but if an insect were programmed to terminate diapause in response to a few days of warmth, the warmth of autumn or a spell of warm weather in winter would trigger the premature termination of diapause, thereby dooming the insect to death when the usual cold of winter returns.

How can this dilemma be resolved; how can insects cope with the unpredictability of the weather? Many insects—including the cecropia moth, a colorful and beautiful creature with a five-inch wing span, whose diapause-terminating mechanisms have been the object of much of the research I have done with Jim Sternburg[15]—avoid the deadly trap of prematurely terminating diapause by responding to a period of warm weather only after they have been exposed to an adequately long period of cold weather. We know that cecropia pupae, the diapausing stage, keep track of the amount of coldness they have been exposed to—much as a meteorologist calculates and keeps a running total of heating degree days in winter. Under natural conditions cecropia pupae terminate diapause, shed the pupal skin, and emerge from their cocoons as moths in response to a period of warm weather, but only after they have been sufficiently chilled.

If you can find a few cecropia cocoons in autumn, you can, if you are curious and have an inquiring mind, demonstrate to yourself the effect of chilling on the diapausing pupae in the cocoons. First, check the cocoons to see if the pupae are alive by gently shaking them next to your ear. If a cocoon is heavy and you hear a solid thud, it probably contains a live pupa. Then put a few of the cocoons in a warm place in your home and put an equal number in your refrigerator. After ten weeks, take the cocoons out of the refrigerator and put them in the warm place right next to the other cocoons. After about three weeks, the moths will, largely in synchrony, emerge from the chilled cocoons. But the other cocoons will just sit there; a few moths may emerge weeks or months later, but most will die without metamorphosing to the adult stage.

Like cecropia pupae, bagworm eggs do not hatch until after they have been exposed to cold temperatures for several weeks. Bob Morden and I[16] determined the effect of increasing durations of natural chilling on the hatching of bagworm eggs. We kept eggs in their bags outdoors in screened cages, where they were exposed to the vicissitudes of the weather from the time they were laid in early fall until mid-May, when all of them were ready to hatch. At regular intervals we determined the progress of diapause termination by transferring some of the eggs to our laboratory and holding them at a constant temperature of 84° F, a temperature conducive to hatching. We found that as the duration of natural chilling increased, the eggs were increasingly conditioned to hatch as they normally do in spring. Eggs transferred in September, before they had experienced a significant amount of chilling, suffered 75 percent mortality and hatching was not synchronized: the first egg hatched after 68 days and a few more caterpillars hatched sporadically during the following 70 days. As the duration of outdoor exposure increased, more and more caterpillars survived and hatching steadily became more synchronized. The last group of eggs, transferred 36 weeks later in May, when bagworm eggs normally hatch in nature, had only the 11 percent mortality that is usual for this species and hatched as a tightly synchronized group over a span of only eight days.

During summer the bagworms in the same area grow at about the same rate. Thus, the synchronization that begins when the eggs hatch is fairly well maintained until the bagworms are mature and ready to mate. Tight synchronization is essential to the bagworm and other insects because the more individuals that are present at the same time, the more likely an individual will be to find a mate and reproduce.

But some insects, as I wrote in 1978,[17] hedge their bets by having offspring that terminate diapause and begin to develop at different times of the year or even in different years. Every cecropia female, for example, produces some opportunistic offspring that, after being sufficiently chilled, terminate their pupal diapause and begin to develop with the first warmth of spring. They emerge as adults in May. The same female also produces a larger group of offspring that terminate diapause only after they have experienced a long period of warmth after chilling. They emerge about a month later. Some of the early group may, for example, survive in a year with a severe summer drought that kills the late group, but some of the late group may survive in a year with a late freeze that kills the early group. Many species of plants have a similar bet-hedging strategy, producing seeds with short- and long-term dormancies on the same plant. Khidir Hilu and J. M. J. de Wet[18] of the Crop Evolution Laboratory of the University of Illinois observed, "The long-term dormant grains [seeds] insure population survival when seedlings that germinated immediately after conditions became favorable are lost through environmental catastrophes."

Insects are faced not only with surviving potentially lethal climatic and other physical aspects of the environment, but are also beleaguered by a host of animals, among them birds, that would like nothing better than to make a meal of them. Insects have evolved many ways of escaping predators, but one of the most common ways is camouflage, which helps them to escape notice. Another way is to sting or be poisonous to eat but not deadly. Predators soon learn to spurn these insects, which are usually conspicuously, warningly, colored and easily recognized. Astonishingly enough, some innocuous and eminently edible insects escape predators by bluffing, by mimicking the appearance and often even the behavior of stinging or toxic insects. Our examination of the black swallowtail butterfly will acquaint you with all of these ways of not becoming a meal for a bird or some other predator.

9

ESCAPING PREDATORS BY DECEPTION

Black Swallowtail Butterfly

The black swallowtail butterfly and every other animal is "preoccupied" with surviving until it can produce the offspring that will pass its genes on to future generations. If it is to survive it must eat, but it must also avoid becoming a meal for some other creature. Insects elude insectivorous birds and other animals in many ways—some quite extraordinary. Many simply try to escape, as does the startled house fly that so quickly flies off or the cockroach that scurries to safety. Many insects do not flee but stay put, escaping notice by resembling an inedible object or by freezing in place and depending upon their camouflage to make them invisible. Some, such as ants, bees, and wasps, defend themselves by biting or stinging. Other insects are poisonous if eaten, and they, as do most stinging insects, warn away would-be predators, which perceive colors very well, with bright, conspicuous colors. Some insects that cannot sting and are not poisonous elude predators by bluffing, by mimicking a stinging or poisonous insect—as a fly banded with yellow and black resembles a wasp and the orange and black viceroy butterfly resembles the toxic monarch. The black swallowtail is protected in one or more of these ways in each of the four stages of its development. But these protective measures can be fully understood only in the context of an insect's, in this case the black swallowtail's, life history, lifestyle, and feeding habits.

The black swallowtail (*Papilio polyxenes*) begins life as an egg that its mother laid on a species of plant which she instinctively knew would be accepted as food by her caterpillar offspring. On sunny days, usually between 10:00 AM and 3:00 PM, according to Paul Opler and George Krizek,[1] females make cruising flights just above short-growing vegetation. When they find an acceptable host plant they dip down briefly to place a single yellow egg near the tip of one or just a few of its young leaves. With just a few exceptions, the only plants that black swallowtail caterpillars will feed on are members of the carrot family (Apiaceae). The black swallowtail caterpillar has, as do all larvae of butterflies and moths, chewing mouthparts with strong mandibles for snipping off and then masticating pieces of leaves. Among their host plants are several native species such as cowbane, water hemlock, and water-pennywort. They have become particularly fond of some introduced weeds: wild parsnip, wild carrot (Queen Anne's lace), and the infamous poison hemlock—which was used to kill Socrates when he was found guilty of "neglect of the gods." Sometimes these caterpillars make pests of themselves by munching on and sometimes seriously defoliating crop plants: cultivated forms of wild carrot and wild parsnip, and dill, parsley, celery, caraway, and fennel.

When the caterpillars are ready to molt to the pupal stage, they void their gut and then pupate, attaching themselves to the host plant, a nearby dead plant, a wooden post, or to the bark of a tree or shrub. The pupa, usually referred to as the chrysalis by butterfly collectors and watchers, is naked (not enclosed in a cocoon) as are those of almost all butterflies. (In the order Lepidoptera it is mainly moths that spin silken cocoons.) Hooks at the tip of the pupa's abdomen attach it to a small pad of silk on the surface of its pupation site. It, as do the pupae of all swallowtails and some other butterflies, leans back at an angle of about 30 degrees from its more or less vertical support, embraced and supported by a thin girdle of silk, just as a worker climbing a wooden telephone pole leans back in his leather safety belt. Because butterflies secrete silk only in the caterpillar stage, the girdle and the pad have to be spun by the caterpillar just before it molts to the pupal stage. This last larval molt is an acrobatic tour de force. With considerable care, the caterpillar sheds its skin while it is within its girdle and the fleshy legs at the tip of its abdomen cling to the pad. It must still be supported by its girdle after the molt and must retain its grip on the pad by shifting its hold from the now-molted abdominal legs to the hooks at the tip of the pupa's abdomen.

Black swallowtails have two or three generations per year, but the final generation spends the winter as a pupa in diapause. The adult that eventually emerges from the pupa is very different from a caterpillar. Its mouthparts, adapted for sucking nectar and other liquids, consist of a long, thin tongue much like a soda straw, that, when not in use, is coiled up under the head like a watch spring. Both sexes drink sugary nectar to provide the energy that fuels their bustling lifestyle, taking it from the blossoms of many different kinds of plants; among their favorites are red clover, bee balm, and several species of thistles and milkweeds. Black swallowtails, powerful fliers, have two pairs of broad, colorful wings that are actually transparent but are covered with tiny, opaque, colored scales that overlap like the shingles on a roof. (These scales are the colored powder that rubs off on your fingers when you handle a butterfly.) Its long, slender legs are its landing gear.

The sluggish, crawling, wormlike caterpillar is designed for eating and growing with maximum efficiency. It has been referred to as an eating machine, and the adult, highly mobile and designed for locating a mate and distributing eggs, has been referred to as a sex machine.

Males and females locate mates when, in the middle of the day, both orient to an open area at some recognizable feature of the landscape, in some areas at the top of a hill, but in flat areas, as told to me by my friend and colleague Jim Sternburg—at a place that is modestly but noticeably elevated above its surroundings—a behavior not unlike women and men congregating at a singles bar. The male gets an unobstructed view of his surroundings by perching atop a plant that rises above the nearby low vegetation. Beginning at 9:00 in the morning and for several hours thereafter, he waits patiently, making only occasional short flights. When a female appears, he sallies out to intercept her. The coupled pair then descend to a shady spot where the male inserts a package of sperm into his mate's vagina. He has thereby accomplished his ultimate goal, perpetuating his genes, if all goes well with the female.

As are virtually all other insects, black swallowtails in all their life stages are food for a host of predators including: many insects such as ants, carnivorous ground beetles, and praying mantises; lizards and small snakes; various birds; and even mammals such as deer mice and skunks. Black swallowtails have several ways of eluding or even warding off predators. Their eggs may in some way be protected, but I know of no published observations or experiments which show that they are. But Jim Sternburg has an idea, as yet untested, that visually oriented predators may pass up yellow eggs on green leaves—perhaps including eggs of the black swallowtail—because they resemble the yellow disease spots that are commonly seen on leaves. Many insects lay yellow eggs while others lay green eggs. It would be very interesting to find out if the host plants of insects that lay yellow rather than green eggs are more likely than other plants to be marked with yellow disease spots.

There is little doubt that the larvae have protections in all of their instars. (Instars, substages of the larval stage, are demarcated by the larval molts.) Black swallowtail caterpillars have four instars separated by three molts. The first two instars resemble bird droppings, which are certainly of no interest to a bird or some other insect eater searching for food. These young caterpillars are predominantly black at both ends but have a broad, white, saddle-shaped band with irregular borders marking one end. The black represents the feces (undigested food), and the white represents semisolid, white uric acid (the urine), which in birds is combined with the feces. Caterpillars of several of the black swallowtail's relatives also mimic bird droppings, among them the tiger, giant, and the rare Schaus's swallowtail of the Florida Keys. All over the world there are spiders, caterpillars, and even adult moths that look like bird droppings. In a letter to Professor E. B. Poulton of Oxford University, Colonel A. Newnham graphically describes a bird-dropping caterpillar of India.

> I came across the larva in question in the month of August or September 1892, at Ahmadabad on a bush of Salvadora. . . . I was stretching across to collect a beetle and in withdrawing my hand nearly touched what I took to be the disgusting excreta of a crow. Then to my astonishment I saw it was a caterpillar half-hanging, half-lying limply down a leaf. [A] thing that struck me was the skill with which the colouring rendered the varying surfaces, the dried portion at the top, then the main portion, moist, viscid, soft, and the glistening globule at the end. A skilled artist working with all materials at his command could not have done it better.[2]

The last two instars—possibly because they are too large to pose as bird droppings on the small leaves that many of their host plants have—are brightly patterned with broad black bands, spotted with yellow dots, on a background of pale green. Opler and Krizek believe that this pattern is cryptic—in other words, that it serves to camouflage the caterpillar.[3] May Berenbaum[4] has argued that it is cryptic when seen from a distance but blatantly conspicuous when seen from up close. But why would a caterpillar, a tender morsel for a bird, "choose" to be conspicuous at any distance?

There are two different answers to this question, one or the other of which applies to most, if not all, conspicuous insects. One answer is that an insect that stings, is poisonous to eat, or is protected in some other way may save itself by advertising its noxiousness, by means of warning coloration, to predators that will not attack it because they remember a previous unpleasant encounter with one of its kind. David Evans and I[5] showed that such an unpleasant memory can persist from one summer to the next in red-winged blackbirds and common grackles. There is a strong evolutionary tendency for noxious insects of different species to adopt similar warning patterns, in other words, to converge on the same pattern. Fritz Müller[6] pointed out that no matter how noxious they are, a few individuals will inevitably be killed in the process of educating and perhaps reeducating predators. If two or more species use the same "advertising logo," there will be an economy for the members of all these species, because the mortality will be shared among a larger group of individuals. Note, for example, that most bees and wasps have similar color patterns of black and orange or yellow bands. Noxious insects that converge on the same warning signals are known as Müllerian mimics—not to be confused with the type of mimicry discussed next.

The second explanation for conspicuousness is that some harmless insects are bluffers that deceive predators by adopting color patterns that are similar to or sometimes almost identical to those of insects that are truly poisonous or otherwise noxious. This phenomenon is sometimes known as false warning coloration but is usually called Batesian mimicry after its discoverer, Henry Bates.

But now back to the question of why black swallowtail caterpillars should be conspicuous only when viewed from close up. Berenbaum[7] suggests, and I think quite rightly so, that this caterpillar is a Batesian mimic of the similarly colored and patterned caterpillar stage of the monarch butterfly. Monarchs, the models for the mimic, feed on milkweeds that contain highly toxic compounds and sequester these compounds in their bodies. It is widely known that adult monarchs are toxic to birds, and experimental studies have shown that they are also noxious to birds in the caterpillar stage. Fortunately these compounds do not usually kill the birds, because they cause them to vomit before they have absorbed a lethal dose. The system of toxins and warning coloration is effective not because it kills birds, but because it creates a population of birds that have learned to heed warning colors. In this way, the warningly colored insects make their "neighborhood" safe for themselves.

The hypothesis that black swallowtail caterpillars, presumably edible, are mimics of toxic monarch caterpillars predicts that naive birds will eat black swallowtail caterpillars but will refuse to eat them after they have had a bad experience with a monarch caterpillar. This prediction can be tested by doing the obvious experiment, "asking" birds themselves to choose between monarch and black swallowtail caterpillars. The results will either support or falsify the Batesian mimicry hypothesis. This has not been done, but a similar experiment with another species, carried out by W. Windecker in 1939,[8] is succinctly described in Malcolm Edmunds's *Defence in Animals*.[9] Windecker found that birds will not eat the yellow and black caterpillars of the cinnabar moth. They picked them up but quickly dropped them and then vigorously wiped their beaks. The noxiousness of these caterpillars resides in the hairs on their body. This is demonstrated by the finding that birds will eat caterpillars that have been shaved. He found that birds which have had an experience with caterpillars with their hairs intact will thereafter reject other yellow and black insects.

Swallowtail caterpillars also have a chemical defense that is known to fend off ants and that probably also drives away other small predaceous as well as parasitic insects. They have an organ, shaped like the forked tongue of a snake, that can be everted and withdrawn from a small slit just behind the head. When this organ, the osmeterium (source of an odor in Greek), is extruded, it gives off a secretion with a characteristic intense odor that is not necessarily offensive to the human nose. Thomas Eisner and Yvonne Meinwald[10] showed that the osmeterium of a close European relative of the black swallowtail is an effective defense against ants. They wrote that as long as the ants attacked in small numbers, the caterpillars easily repelled them. When bitten by an ant, a caterpillar "would instantly revolve its front end so as to wipe its extruded osmeterium against the assailant, and the ant, visibly contaminated with secretion, would promptly flee, pausing occasionally along the way to cleanse itself vigorously."

Black swallowtail pupae are likely to escape the notice of predators because they are camouflaged and blend in with their surroundings. Pupae of the fall generation, which do not emerge as adults until the following spring, are always brown and blend in with the dead plants of winter. Those of the summer generations may turn either brown or green. They will be brown if they choose to pupate on the trunk of a tree, a fence post, or some other dark support, but when the caterpillar attaches itself to a green plant, the pupa will be green. But it is not quite that simple. David West and Wade Hazel[11] found that in summer pupae may be green if they are on a very slender brown stem. They postulated that these pupae escape notice because they resemble leaves in a setting in which "green patches"—leaves—are common components of the background.

How much protection does its camouflage actually confer on a pupa? In the field Hazel and colleagues[12] glued brown and green laboratory-reared pupae to brown supports such as tree trunks or the stems of shrubs or to green supports, the stems of herbaceous plants. Some of the 297 pupae used in their experiment were placed on matching backgrounds and others on contrasting backgrounds. They checked the pupae daily and found, as might be expected, that those on matching backgrounds were more likely to escape predation than were those on mismatched backgrounds.

The most widely known and one of the best-understood cases of camouflage is that of the peppered moth—elucidated mainly by the work of H. B. D. Kettlewell[13] of Oxford University. Before the Industrial Revolution, peppered moths were light in color and inconspicuous on the light-colored bark of trees. Their camouflage was further enhanced because their wings and bodies were patterned to resemble lichens that grew on the bark.

But the forests in which these moths live changed after the Industrial Revolution began with the invention of the steam engine in the late eighteenth century. As coal-burning factories proliferated in the nineteenth century, smoke pollution stained black the trunks of trees in woodlands near factory towns and killed the lichens that grew on their bark. The light-colored moths were no longer camouflaged when they sat on these blackened tree trunks. Only black peppered moths, which were not known to exist in the eighteenth century, would have been likely to escape notice on the smoked-stained bark.

But a black peppered moth, possibly a recent mutant, was found in a previously all-light population near the industrial city of Manchester in 1848. The better-camouflaged black form of this moth then increased as the population of the light form decreased. By 1898 only 50 years later, about 95 percent of the population of peppered moths near Manchester were black. The same thing happened in and around factory towns all over Britain, northern Europe, and, with other species of insects, in North America. Although some black mutants may have appeared in areas where there was no air pollution, natural selection soon eliminated them, and populations of peppered moths in clean areas continued to consist of light-colored individuals. This story has an interesting sequel. As Britain makes progress in cleaning up air pollution, the trunks of once-stained trees are reverting to their original light color and lichens again grow on their bark. This favors the small minority of light-colored moths that survived in these areas, and they are now supplanting the black moths.

But why the difference in survival? An obvious possibility is that birds that scan tree trunks for prey are most likely to find moths that do not match the color of the bark on which they rest. Kettlewell and his colleagues[14] eventually showed this explanation to be correct, and others found that the black moths have the additional advantage of being better able than light moths to cope with the physiological stress caused by smoke pollution. Two tactics were then used to show that birds are important in the selection process. First, moths were released on both light and dark tree trunks and later recaptured at lights, to which they are strongly attracted, or at traps baited with pheromone-releasing females. Moths were more likely to survive if they matched their backgrounds. More moths released on bark of the "right" color were recaptured than those released on bark of the "wrong" color. The next step was to determine if birds were indeed responsible. Moths were placed on matching or contrasting tree trunks and watched from a blind. Birds were often seen capturing those that did not match the color of the bark on which they sat, but they very seldom noticed moths that did match their background.

The adult female, but not the male, black swallowtail in the butterfly stage, is one of several insects that were, long before there was evidence to prove the point, thought to be Batesian mimics of the pipevine swallowtail. The pipevine swallowtail, the model, is noxious because it feeds on the notoriously toxic plants of the genus *Aristolochia*, such as the ornamental Dutchman's pipe and the wild Virginia snakeroot. It is blatantly warningly colored, with wings that are almost totally black on the upperside and black with several large orange spots on the underside. The other insects that mimic the pipevine swallowtail—more or less faithfully—are four butterflies: the black phase of the female tiger swallowtail, the female Diana fritillary, and both sexes of the spice bush swallowtail and the red-spotted purple; and one moth, a recent addition to this list, the day-flying male of promethea, which, along with the beautiful but nonmimetic cecropia and luna moths, is one of the giant silkworms of North America.

Why do only the females of some of these species mimic the pipevine swallowtail? Why aren't the males protected against predators by mimicry? No one knows, but there are two reasonable hypotheses. One is that females gain more from being mimetic than do males, because females are more exposed to predators, perhaps because they are more active. There is little supporting evidence from butterflies, but the reversed sex-limited mimicry of promethea suggests that this hypothesis may well be correct. The day-flying males, exposed to visually oriented birds, are mimetic, but the night-flying females, not exposed to these birds, are not mimetic. The other hypothesis is that female butterflies are more likely to accept courting males with the ancestral color pattern of their own species than males with the more recently evolved mimetic pattern. The idea is that any benefit the male gains from mimicry would be more than offset by his decreased ability to attract and inseminate females, but there is little evidence to support this hypothesis.

In 1958 Jane Van Zandt Brower[15] published the results of laboratory experiments with butterflies and insect-eating birds that give strong support to the Batesian mimicry hypothesis. The birds were Florida scrub jays: some naive control birds that had not been exposed to the toxic model, and other experienced ones that had already learned to reject the model. The nontoxic butterflies were spice bush swallowtails, female black swallowtails, and black phase female tiger swallowtails. The results were remarkable. An experiment comparing the reactions of the birds to pipevine swallowtails and nontoxic but mimetic black swallowtails gave particularly convincing results. All of the naive control birds pecked at, killed, or ate all of the black swallowtails offered to them. But 12 out of 13 experienced birds, well aware of the models' toxicity, did not even touch the mimics. These results cannot be easily dismissed. A statistical analysis showed that there is only one chance in a thousand that they could have occurred by random chance.

Lincoln Brower[16] went a step further: he raised some monarchs on a species of milkweed that does not contain the digitalis-like substances that make monarchs noxious, and he raised other monarchs on milkweeds that do contain them. Blue jays that had been held in captivity long enough to forget any previous experience with a toxic monarch readily ate monarchs raised on poison-free plants. They continued to eat nontoxic monarchs as long as they were offered. Jays that were offered monarchs that had been raised on toxin-containing plants ate them but quickly showed obvious signs of distress, erecting their crests and fluffing out their feathers. Then they became ill and vomited. After that one experience, they refused to eat either toxic or nontoxic monarchs, and some of them retched when they so much as saw a monarch.

The close resemblance of the viceroy butterfly to the toxic monarch butterfly has long been cited as an outstanding example of Batesian mimicry, but recent studies show that naive birds of some species will not eat viceroys, suggesting that they may be Müllerian rather than Batesian mimics. But there are numerous indisputable examples of Batesian mimicry, such as the swallowtail mimetic complex and numerous examples of wasp- or bee-mimicking flies that are readily eaten by naive birds but not by experienced birds.

But some biologists pointed out, and quite rightly so, that laboratory experiments do not necessarily reflect what happens in nature. It would, of course, be impossible to make enough observations of wild birds attacking or rejecting wild butterflies to make any definitive pronouncements. What to do? Lincoln Brower, then at Amherst College in Massachusetts, came up with a brilliant idea that probably comes as close as possible to measuring mimetic advantage in nature. He proposed releasing and recapturing male promethea moths that had been painted to look either like noxious butterflies or palatable butterflies. This is not as far-fetched as it sounds. Male promethea fly during the day as do butterflies; they are about the same size as swallowtails, they look and fly like butterflies, and their wings are almost totally black. When they are in flight their resemblance to black pipevine swallowtails or one of their mimics is uncanny. On several occasions Jim Sternburg and I pursued and netted promethea males that we had been certain were swallowtails. On a sunny day I sat in my car in my driveway waiting for my wife. A big black promethea fluttered across my lawn. I was certain it was a swallowtail until it tried to beat its way into the screened porch, where I had several virgin promethea females in cages.

This male promethea was obviously responding to a sex attractant pheromone released by the virgin females. (Once a female has been inseminated she never releases this pheromone again.) In nature, virgin female prometheas, which with their inconspicuous brown, tan, and reddish wings look nothing like a swallowtail, hide in foliage, where they are relatively safe from diurnal predators, and release their sex attractant pheromone from about 2:00 PM or earlier on cloudy days, until shortly before sunset. Because females fly only at night, when they lay their eggs, they have nothing to fear from diurnal birds and need not be mimetic. Some males fly in the morning, but most fly in the afternoon while the females are releasing the sex attractant pheromone. Unlike females, the hapless males are exposed to diurnal insectivorous birds, and have only their mimicry of the pipevine swallowtail for protection.

Now back to Brower's idea. The black wings of a male promethea are a blank "canvas" on which many different patterns of various colors can be painted. Painted promethea males released in nature were exposed, sometimes for two days or more, to wild, diurnal, insect-eating birds until they were recaptured. One of the main advantages of using male prometheas is that they can be recaptured with little effort in traps baited with pheromone-releasing virgin females. Butterflies, on the other hand, would have to be hunted down one by one with an insect net, because neither sex releases a pheromone that attracts the other sex from a distance. Needless to say, a release and recapture experiment with butterflies would be at best extremely difficult and probably impossible.

Brower and his group did their release and recapture experiments in Trinidad, painting some moths to resemble colorful butterflies known to be noxious, while others, the presumed edible controls, had black paint applied to their black wings. This did not change their resemblance to black-colored toxic butterflies such as the pipevine swallowtail or three other related toxic swallowtails, that are also largely black and occur on Trinidad. After several years of experimentation with this system, none of their results constituted a convincing demonstration of the efficacy of Batesian mimicry.

In summarizing their release and recapture experiments, Laurence Cook, Lincoln Brower, and John Alcock[17] state: "Taking all the evidence over the four years there is no significant advantage to either mimic or control moths . . . and no heterogeneity between years or between new and old sites, and perhaps it should be concluded that under wild conditions no clear selective differential can be demonstrated with the *promethea* moth mimicry system." They felt that the painted mimics may at first have had an advantage over the controls, but that this was rapidly converted to a disadvantage as birds became aware of the experimental insects. Throughout their experiments the Brower group made the tacit assumption that the normal appearance of male promethea does not tend to ward off predation by Trinidad birds. Jim Sternburg and I[18] argued that this assumption is not justified, and that their data can be reinterpreted as a convincing demonstration of the efficacy of Batesian mimicry.

The crux of the matter is, as I have explained, that, unknowingly, Brower and his coworkers had compared the survival rates of moths painted to resemble colorful toxic Trinidadian butterflies with the presumed control group.[19] But the presumed controls actually resembled another group of toxic butterflies on Trinidad, three species closely related to the pipevine swallowtail, the natural model of promethea in the United States. The appearance of the black prometheas was not changed by weighting them with black paint, although this was a control for the effect of applying paint to the moths. But the black-painted prometheas were certainly not obviously edible controls that had no resemblance to toxic butterflies, as the Brower group had supposed. They were definitely edible but they did look like toxic butterflies, the Trinidadian relatives of the pipevine swallowtail. The appropriate controls would have been promethea males painted to resemble an edible butterfly.

In central Illinois, Michael Jeffords, Jim Sternburg, and I[20] did release and recapture experiments similar to those done by the Brower group on Trinidad. The difference was that we compared male prometheas with black paint on their black wings (mimics of the pipevine swallowtail) with moths painted with yellow stripes (caricatures of the yellow form of the edible tiger swallowtail). On many days between late June and early September we released equal numbers of each painted type in the middle of a mile-wide circle of seven traps baited with virgin female prometheas that surrounded an area of the University of Illinois's Robert Allerton Park that consisted mainly of mature forest but also included small areas of restored prairie and young forest.

We recaptured 40 percent of the 436 moths that we released. All had had to run the gauntlet of insectivorous birds for at least a half mile—probably often much more—before they entered a trap. The results are a convincing demonstration of the efficacy of Batesian mimicry. We recaptured about 47 percent of the black-painted moths but only 34 percent of the yellow-painted ones, statistically a very significant difference. Furthermore, many of the recaptured yellow-painted moths, but very few of the others, had injuries to their wings, occasionally the clear imprint of a beak, that had obviously been caused by birds that had snapped at them but had not been able to hold them.

The lives of all animals are governed by three imperatives: they must, as we have seen, reproduce themselves, and in order to do so, must avoid being eaten, and must themselves eat and grow to sexual maturity. Most plant-feeding insects are picky eaters that will feed on only one species of plant or, much more often, on just a few related species of the same family. These fussy feeders are said to be "host specific." Among them is the cabbage white butterfly that you will meet next. With a few notable and revealing exceptions, it will eat only the leaves of cabbage and other plants of the mustard family, such as kale, collards, cauliflower, brussels sprouts, and turnips. We will next consider why insects are host-specific as well as how they manage to find and recognize their proper food plants.

10 WHY INSECTS ARE SUCH PICKY EATERS

Cabbage White Butterfly

Imported cabbage worms confined in a laboratory greedily eat and thrive on leaves of various plants of the mustard family. But when provided only leaves of some other kind of plant, perhaps spinach, lettuce, or tomato, they may take a few little test nibbles, but then starve to death without feeding. But in 1910, Dutch entomologist E. Verschaffelt[1] induced caterpillars of this species to eat filter paper, which has absolutely no nutritional value for them. The reason for these bizarre and seemingly contradictory results will, believe it or not, make sense as you read on.

The cabbage white butterfly (*Pieris rapae*), usually called the imported cabbage worm in North America, is an invader from Europe that is familiar to gardeners who grow cabbage, brussels sprouts, kale, turnips, or other plants of the mustard family (*Cruciferae*). The green, velvety caterpillars, as much as an inch or more long, riddle leaves with large irregularly shaped holes. When numerous, they will do costly injury to a crop, destroying so much leaf tissue that the growth

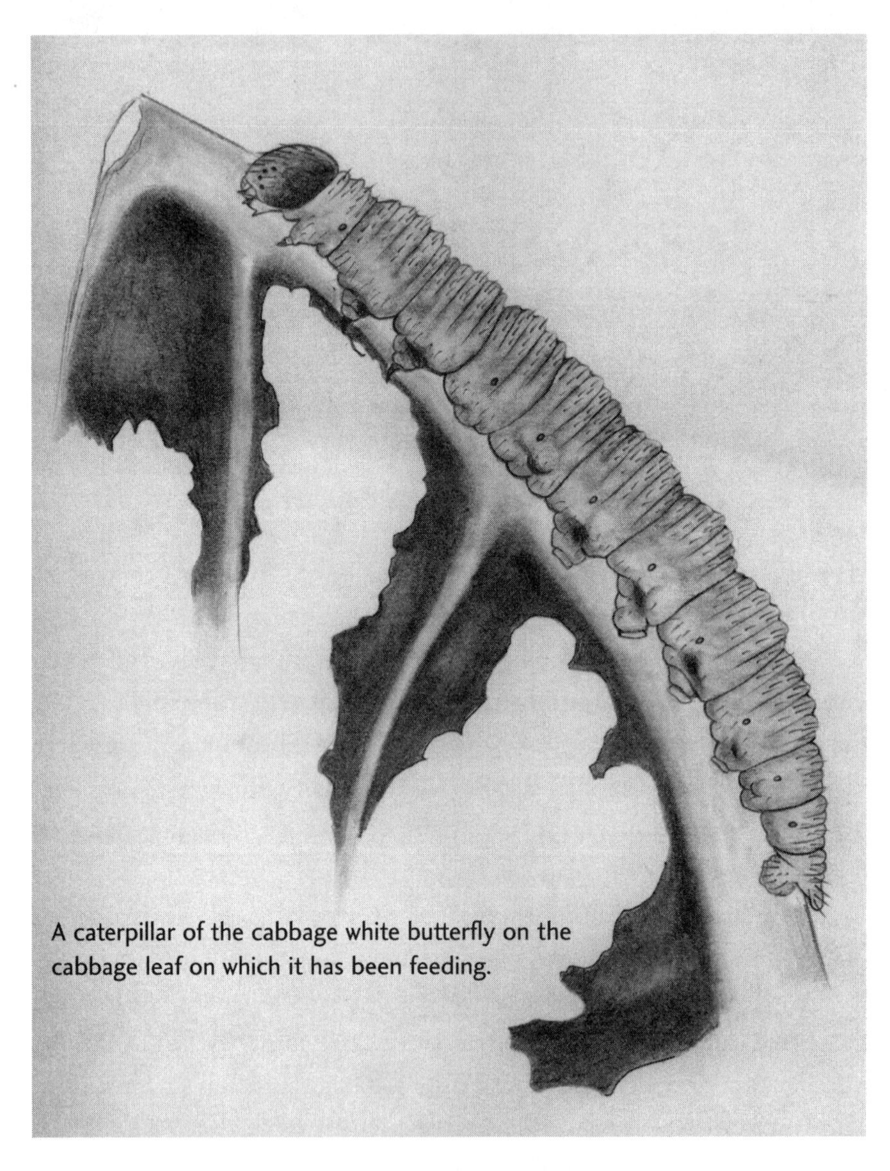

A caterpillar of the cabbage white butterfly on the cabbage leaf on which it has been feeding.

of the plants is greatly diminished; kale, collard greens, and other leafy crucifers are rendered unfit for consumption; and the heads of cauliflower and cabbage are likely to be stunted or not form at all.

When full-grown, the caterpillars pupate on the plant or some other nearby support. The naked pupa, or chrysalis, is suspended as is that of a black swallowtail. Hooks on its tail end are embedded in a button of silk it spun on the support, and, leaning back at an angle, it is further supported by an embracing girdle of silk. The adult that emerges a week or two later is the familiar white butterfly, marked with a few black spots on the wings, that we see early in spring and, since there are three or more generations per year, is also common throughout the summer. The dark yellow eggs, several hundred per female, are laid singly on the underside of a leaf. Pupae of the last generation spend the winter in diapause attached to a weed stalk, fence post, or some other permanent support.

If not confused by an experimenter wielding filter paper, most plant-feeding insects are surprisingly fussy about what they eat. Of the 400,000 currently known plant-feeding insects, about 80 percent, 320,000 species, are, as Louis Schoonhoven and coauthors[2] explain, more or less host-specific, that is, they feed on plants of only one family or a few closely related families. According to Niklas Janz and coauthors,[3] even strict specialists that feed on only one plant or a few plants of one family greatly outnumber generalists that will feed on many different kinds of unrelated plants. For example, silkworms refuse to feed on the leaves of plants other than the white mulberry or a few of its close relatives; tomato hornworms and Colorado potato beetles feed only on plants of the nightshade family; monarch caterpillars only on milkweeds; bean leaf beetles only on plants of the pea family; elm leaf beetles only on elms; and squash bugs suck sap only from members of the cucumber family.

This is in sharp contrast to the plant-feeding mammals, which, with only two exceptions, include in their diets many different kinds of plants of many families. In other words, they are polyphagous, generalist feeders, as are a minority, about 20 percent, of the plant-feeding insects. We ourselves and many of our primate relatives eat the leaves, stems, tubers, fruits, or seeds of a wide variety of plants. A few of the many eaten by people are beans, squash, spinach, lettuce, wheat, cabbage, celery, apples, and oranges—each just one species of just one family of plants that also includes anywhere from a few to many other species eaten by humans. The two exceptions to the usual polyphagy of mammals are the charming koala of Australia, which eats only the leaves of eucalyptus trees, and the giant panda of China, which is reputed to eat only bamboo.

There are many different kinds of plants—300,000 in the world, 2,400 just in Illinois, and almost 1,500 of 136 different families in just the 12 counties of Illinois's Sangamon River basin. Why do so many insects, surrounded by a multitude of potential food plants, pass up so many plants that are successfully utilized by other insects? In other words, how do they benefit by being host-specific, by feeding on only a few closely related plants? Specialization has its obvious advantages. Few people, for example, try to do everything; they become specialists, skillful carpenters, librarians, teachers, or plumbers.

And so it is with host-specific insects. Although the plants in one family are likely to be similar to each other, more distantly related plants generally differ from each other in many ways that affect plant-feeding insects: in their size and growth form, their biochemical and physical defenses against marauding insects, in the microclimate of their habitat, and in the seasonal timing of events in their life cycle, such as germination, flowering, and the release of seeds. Consequently, insects evolve so as to best cope with their own host plant by adapting to the attributes peculiar to it.

The benefits that specialization confers on plant-feeding insects are probably as many and varied as are the multitudinous challenges they face as they strive to grow, survive, and produce offspring. An insect may escape competition from other plant feeders by adapting to cold temperatures so that it can feed on an early spring plant that is utilized by relatively few competing insects. Some insects may be tied to a specific host because the camouflage that protects them from predators is geared to the color, leaf size and shape, and other visual attributes of its host plant. Most often host-specific insects benefit by limiting themselves to plants whose particular physical and chemical defenses they can breach. They may, for example, have evolved strong jaws to bite into a tough-walled leaf or fruit, a way of avoiding thorns or sharp bristles, or the physiological ability to tolerate or detoxify a plant's biochemical defense. Monarch butterflies, for one, feed with impunity on milkweed plants containing toxins that keep almost all other insects at bay. As have some other insects, they have even turned the tables on the plant by incorporating these toxins into their own bodies as a protection against insect-eating birds. The conspicuous orange and black coloration of the monarch warns away birds that remember the ill effects of having eaten one. Only a few insects feed on milkweeds, but most of those that do, among them a leaf beetle, a sap-sucking bug, and a long-horned beetle, also sequester milkweed toxins and also display orange and black warning colors.

Host-specific insects have an uncanny ability to find and recognize their host plants. With very few exceptions, adult females lay their eggs on the "right" species of plant, which they instinctively "know" will be acceptable food for their offspring. They very seldom make a mistake. For instance, caged tobacco hornworm moths, Robert Yamamoto and Gottfried Fraenkel[4] found, do not hesitate to lay eggs on potted tomato, tobacco, or other plants of the nightshade family, but they will not lay even one egg on any other kind of plant.

Immature insects, larvae or nymphs, even upon hatching from the egg can recognize and accept or, on the rare occasions when an egg-laying mother made a mistake, reject the plant on which they find themselves. This ability, apparently inherent in all host-specific insects, seems superfluous because of the mother's almost infallible ability to place her eggs on a plant that will be acceptable to her off-spring. But the capacity to recognize the proper host plant will be useful to larvae or nymphs later in their lives if they need to find another host plant because they are somehow separated from their natal host plant. A caterpillar may, for example, drop to the ground to escape an approaching parasite or predator. An insect may for some good reason have to move from one host plant to another, perhaps because just one plant is too small to satisfy its appetite, or perhaps to move away from leaves it has damaged, which are visual cues that can reveal its presence to a hungry bird.

Host-specific insects are often said to be the plant taxonomists of the insect world, because they can recognize the various species of the plant family they favor and can distinguish them from virtually all other plants. It is also reasonable to think of them as "good biochemists," because they identify plants mainly by the substances that give them their characteristic smell or taste but are not essential to the plants' existence. Vegetables and fruits such as celery, fennel, cherries, and cabbage owe their characteristic odors and flavors to these substances. We value spices such as vanilla and pepper and herbs such as sage, rosemary, and thyme for the odors and flavors of the substances they contain. As we will see, these scents and flavors are those of secondary plant substances, which are not essential to the metabolism and other physiological processes of a plant—as opposed to the essential primary substances, nutrients such as minerals, carbohydrates, and fats.

How did this multitude of secondary plant substances that exists today come into being? They first appeared, and new ones surely continue to appear, as the result of genetic mutations in individual plants. But if a mutation is to survive and spread to the descendants of a mutant, it must pass the muster of natural selection. In other words, it must prove to be useful, thereby increasing the fitness of the individual plants or animals that harbor it. Some secondary substances are favored by natural selection because they have scents that attract pollinating insects to blossoms; the floral scents that delight our noses may be even more pleasant to the insects that are rewarded with nectar when they visit these blossoms. Other secondary substances that arose by mutation were conserved by natural selection because they proved to be biochemical defenses against the enemies of plants. Viruses, bacteria, fungi, and various vertebrates are among these enemies, but I will focus on defenses against plant-feeding insects. Some of these defensive substances are noxious, causing insects or other plant feeders to suffer unpleasant symptoms or even to die. Others are not in themselves noxious, but have characteristic scents or tastes that dissuade an insect from feeding, by warning it of the presence of some other substance that is noxious.

The first significant step in unraveling the complex and fascinating story of the interactions between insects and plants came in 1910, when Verschaffelt demonstrated that caterpillars of the cabbage white butterfly feed only on plants of the mustard family and not on plants of other families, but with a few revealing exceptions, the ones that prove the rule. Among the exceptions are two common garden flowers, the nasturium of the watercress family and the spider flower (*Cleome*) of the caper family. When Verschaffelt bruised leaves of these plants, he found that they had the same pungent odor as cabbage and the other crucifers. It was already known that the pungent odors of crucifers and plants of the caper and watercress families are caused by chemicals called mustard oil glucosides, also known as glucosinolates. He hypothesized that these chemicals are the attractants that prompt cabbage white caterpillars and quite a few other insects to feed on crucifers. The results of his clever experiments supported this hypothesis:

> To what extent indeed these insects are attracted by mustard-oils is clear when the leaf of a species not otherwise eaten by them . . . is smeared with a paste or the juice obtained from the leaves of a Crucifer . . . and is offered them as food. It was at once attacked and in a short time devoured. The same occurrence with the leaves of other plants needs no further explanation, though it will easily be understood that every species cannot be used for such experiments . . . [some] remain untouched, doubtless because they contain constituents which are distasteful to the caterpillars.
>
> It is however unnecessary to place the Crucifer-juice on a living leaf . . . [filter paper], for example, which is rejected by . . . *Pieris*-caterpillars

> when dry or moistened with water, I saw eaten with avidity when soaked with [crucifer] juice. A microscopical examination of the excrements showed scarcely any other constituent in it than the matted paper fibres.[5]

Except for accounts of a few inconclusive experiments by others, little was added to our understanding of how host-specific insects manage to recognize their favored plants until the 1953 publication of the results of Asgeir Thorsteinson's seminal research at the University of London.[6] The diamond-back moth, the subject of his research, is a specialist that in the caterpillar stage feeds, like the cabbage white, only on crucifers and a few chemically similar plants. Gottfried Fraenkel,[7] a pioneering student of host plant specificity, remarked that, "Thorsteinson's results entirely corroborated Verschaffelt's . . . and in many respects extended them. . . ."

Thorsteinson demonstrated, as did Verschaffelt with the cabbage white, that diamond-back moth caterpillars specialize in feeding on crucifers. They readily accepted over 30 species of this family and, like the cabbage white, a few noncrucifers that also contain mustard oil glucosides, but rejected all other plants. Then moving forward from where Verschaffelt had left off, Thorsteinson did chemical analyses of crucifers, the first step in proving that, as Verschaffelt had suggested, mustard oil glucosides are the attractants that prompt host-specific insects to feed solely on these plants. Thorsteinson isolated and purified two mustard oil glucosides, sinigrin and sinalbin, from black mustard seeds—and also myrosin, an enzyme that has an essential function in the mechanism by which crucifer-feeding insects recognize their host plants. All of these substances are also present in the leaves of crucifers, not only in their seeds.

He next did the critical experiments which showed that diamond-back moth caterpillars are actually induced to eat by these mustard oil glucosides. They readily fed on the leaves of plants that they would not ordinarily eat, common groundsel of the daisy family and the common garden pea of the pea family, if the leaves were coated with either sinigrin or sinalbin. Thorsteinson went on to incorporate powdered pea leaves and several concentrations of sinigrin in agar gels, which in themselves are tasteless and contain no nutrients. The caterpillars nibbled just a bit on gels that contained pea leaf powder but no sinigrin. However, they ate substantial quantities of the same gels if sinigrin was added—even if the sinigrin content was only 7.8 parts per million parts of gel and powdered pea leaves. Given this result, it seems likely that a leaf containing only a tiny amount of sinigrin will be recognized as an acceptable food by diamond-back moths. They are far more sensitive to sinigrin than are people. Thorsteinson compared his own and a diamond-back caterpillar's ability to taste sinigrin. He was "scarcely able to taste the most concentrated sinigrin solutions tested," which were 1,000 times as concentrated as the lowest concentration that the caterpillars were able to perceive.

Thorsteinson made another discovery whose full significance was at that time not fully understood. He found that caterpillars ate less when the enzyme myrosin was incorporated in a gel containing powdered pea leaves and sinigrin. Because the amount eaten was not decreased by the incorporation of the enzyme inactivated by heat, he concluded that the enzyme was of itself not repellent, but that the active enzyme chemically depleted the amount of sinigrin in the gel.

The essential role of the enzyme myrosin in the chemical defense of crucifers, clarified 50 years later by Andreas Ratzka and coworkers,[8] explains how caterpillars of the diamond-back moth were partially deterred from eating Thorsteinson's pea leaf powder and sinigrin-containing gel when myrosin was added. This is, furthermore, one way in which, as Gottfried Fraenkel,[9] a great insect physiologist, had previously postulated, a plant toxin can become a harmless token feeding stimulus. Sinigrin and other mustard oil glucosides, of little harm in themselves, produce toxic feeding deterrents when a leaf is damaged, as by the chewing of an insect, and they come into contact with myrosin, which had been separately stored in the leaf. Myrosin causes mustard oil glucosides to break down into a variety of products, some of which are toxic to insects not specialized to feed on crucifers and even to crucifer specialists such as diamond-back moths. But this moth, and presumably other crucifer-feeders, are not poisoned because they secrete a special enzyme that prevents myrosin from breaking down mustard oil glucosides, the token feeding stimulants, to toxic substances that deter feeding. Because the myrosin Thorsteinson added to his gel was not separated from the sinigrin in the gel, as it would be in a leaf, the myrosin reacted with some of the sinigrin, producing enough toxic and deterrent breakdown products to slow down the rate of feeding.

Fraenkel's 1959 article, "The Raison d'être of Secondary Plant Substances,"[10] proposed a unifying theory of host plant selection. He argued that since the known nutrient requirements of insects are all similar and that since these nutrients, required by all living things, are contained in the leaves of all plants, it is not likely that host-specific insects recognize their particular food plants by its primary substances, which constitute its nutrient composition. He said that leaf-feeding insects could grow and develop equally well on the leaves of any plant, provided they ate enough of them. Insects, he proposed, recognize and choose to feed on a plant solely in response to its secondary substances, which arose in the first place as defenses against insects. "A host preference," he argued, "arose when a given insect species, by genetic selection, overcame the repellent effect of [such a defense], thereby gaining a new source of food. This led to a situation where further selection produced new species . . . of insects that require the former repellent as an attractant [token feeding stimulant] to induce feeding" (p. 1466).

Fraenkel overstated his case by saying that leaf-feeding insects could grow equally well on any plant, given that they ate enough of it. This did not detract from his basic argument, but it did misleadingly oversimplify the physiological and ecological aspects of the interactions of insects with their host plants. Working on my doctoral dissertation under Fraenkel's guidance, I found that after the amputation of certain of their organs of taste, tobacco hornworms ate some plants not of the nightshade family, plants they would not otherwise have eaten, probably because they could no longer taste deterrent substances in these plants.[11] I later found that hornworms lacking these taste receptors fed as rapaciously on burdock (daisy family) and mullein (snapdragon family) as on tomato (nightshade family), their usual food plant, but gained, respectively, only 74 percent or 44 percent of the weight they gained on tomato. This is probably not because of differences in the *kinds* of nutrients in these plants, but rather because of the differences in the *proportions* in which they are present—in other words, to differences in the balance of nutrients.

In 1965 Paul Ehrlich, an entomologist, and Peter Raven, a botanist, analyzed the known food plant preferences of butterflies, and, on the basis of their findings, restated Fraenkel's proposition in an evolutionary and ecological context.[12] (In the same year, G. E. Hutchinson published a book with the apt title *The Ecological Theater and the Evolutionary Play*.)[13] They, as so concisely summarized by Arthur Weis and May Berenbaum,[14] envisioned a process by which insects evolved their disparate feeding habits: A chance mutation enables a plant to synthesize a novel secondary substance that happens to make it a less suitable food for insects. Freed from its insect enemies, the plant undergoes an evolutionary radiation (divergence) that produces new species. By random chance, an insect has a mutation that makes it resistant to the plant's new defense. Benefiting from this previously unavailable food resource, the insect flourishes and undergoes its own evolutionary radiation. Ehrlich and Raven[15] referred to this process as an example of coevolution, the reciprocal evolution of two organisms or groups of organisms in response to each other.

Put more simply, plants evolve defenses against the insects that eat them. If the insects are to survive, they must switch to other foods or evolve ways to circumvent the plants' defenses. Insects soon come to prefer the plants whose defenses they can circumvent, and eventually evolve the ability to identify them by their characteristic flavors or odors, or both. A plant species whose defense has been breached must then evolve another defense or fall by the wayside. The insects must respond in turn.

The tens of thousands of defensive secondary plant substances that exist today evolved in this manner during the past 300 million years or more, each one a challenge to one or more insects that depend upon plants for food. Today we see around us the survivors of this coevolutionary arms race, hundreds of thousands of plant and animal species that evolved as coevolution drove both plants and animals to become ever more specialized. But thousands, hundreds of thousands, or even more species—both plants and animals—must have gone extinct because they could not keep up in the arms race.

Just as plants have ways of defending themselves against their insect enemies, these same insects have evolved defenses against *their* insect enemies, among them parasitic wasps and predators such as ants and beetles. The chemical defense of the cabbage white caterpillar was discovered, analyzed, and then described in a 2002 issue of the *Proceedings of the National Academy of Sciences* by Scott Smedley and several colleagues,[16] among them Thomas Eisner, long recognized as the leading authority on the chemical defenses of insects.

They found that the caterpillar's back bristles with long glandular hairs are each topped with an oily droplet of a defensive secretion that consists mainly of a mixture of unsaturated lipids (fats). (Smedley and his colleagues named them "mayolenes" in honor of May Berenbaum, "whose visionary research, lectures, and writings on insects have helped spur entomophilia [love of insects] the world over.") Ants that contacted droplets of mayolenes when they approached a caterpillar retreated in obvious distress and licked and otherwise cleaned their bodies for an unusually long time, more than six times longer than ants that had contacted caterpillars from which the mayolenes had been removed.

This continuing struggle between plants and insects is the reason why so few insects are "generalists" that feed on many different kinds of plants, and why the great majority are "host specific." The latter feed on only a few closely related plants—those whose defenses they can breach—and spurn hundreds of thousands of other species, although all of them contain all the nutrients required by them and, with few exceptions, by all living things.

Among the generalist feeders is the corn earworm, which, despite its common name, is by no means host-specific. It is a destructive pest not only of corn, but also of cotton, tobacco, tomato, and vetch, and also feeds on many wild plants. We will next not only consider this insect as a pest, but also find how, in the laboratory, corn earworms can select for themselves a balanced diet from separate nutritional components, protein, a carbohydrate, fats, vitamins, and minerals. This is not only a laboratory curiosity. Insects in nature are faced with the same problems. Quite a few must take in a favorable balance of different plant parts that vary in nutritional composition and some even a favorable mix of plants of two or more species.

11

"NUTRITIONAL WISDOM"

Corn Earworm

It is Saturday and the farmer's market is open; you are there shopping for fresh locally grown vegetables, especially sweet corn. After a winter of making do with ears of aging corn hauled in from California or Florida, you are hungry for some freshly picked sweet corn with plump, juicy kernels. You are in luck. A farmer has brought in a load of corn, the first of the year, that he picked the previous afternoon. It is one of your favorite varieties, the deliciously sweet ivory and gold. You pick up an ear, pull the husk partway back, and find a big, fat caterpillar that has been eating the kernels at the tip of the ear. You toss the ear back and pick out some others that are not inhabited by caterpillars. Those who grow sweet corn in home gardens will be familiar with this insect and take it in stride. The caterpillar—there is usually only one per ear because they are cannibalistic—will not have ruined the ear. It didn't eat very much, only a few kernels at the tip of the ear, and the small damaged part can be chopped off.

This caterpillar, the corn earworm (*Helicoverpa zea*), is the larval stage of a dull-colored moth with a wingspan of about 1.5 inches. Its name is misleading because it is not a specialist that feeds only on corn (maize). Herbivorous insects may be monophagous, willing to feed on only one or a few closely related plants; oligophagous, feeding on a wider variety of plants, often several species of the same family; or polyphagous, feeding on a very wide variety of plants from several different families. The corn earworm is blatantly polyphagous, as indicated by some of its other common names: false tobacco budworm, vetchworm, cotton bollworm, and tomato fruitworm.

When infesting corn, corn earworms prefer to lay their eggs on fresh silks. The eggs are placed one at a time. But a moth may put several on the silks of one ear and several moths may lay eggs on the same silk. I have seen tufts of silk with hundreds of eggs. The egg, glued to a strand of silk, is light yellow and about half the size of the head of a common pin. In warm weather, eggs hatch in as little as two days and the caterpillars immediately burrow deep into the tuft of silks, feed on the silks until they dry out, and then switch to feeding on kernels.

The larvae, write R. A. Blanchard and W. A. Douglas,[1] US Department of Agriculture entomologists, grow very rapidly, attaining full size in, depending upon the temperature, 13 to 28 days after hatching. The full-grown caterpillar, which may be patterned in green, yellow, orange, *or* brown, is robust and about 1.5 inches long. It bores its way out through the husks and makes its way to the soil. "It enters the soil," explain Blanchard and Douglas, "as soon as possible and bores down 1 to 9 inches. . . . The larva then forms a cell . . . and digs a passageway from the cell, almost to the surface of the soil, through which it will escape after it has become a moth. It then returns to the cell, where it transforms into a light-brown pupa about 3/4 inch long."

If the pupa experienced the long days of summer when it was in the larval stage, it soon begins to develop and emerges from the soil in two or three weeks. If it experienced the short days of autumn as a larva, it stops developing and goes into diapause. Throughout most of the corn belt there are three or four generations per year; pupae of the last one are in diapause throughout the winter and do not emerge as adults until the next spring.

Adult corn earworms are nocturnal. They emerge from the pupal cell in the evening, according to Philip Callahan,[2] an agricultural entomologist, about 95 percent of them between the hours of 7 and 11 PM. On succeeding days they become active early in the evening and drink the nectar of various flowers. Females release a sex pheromone into the air to attract a male, and after mating fly in search of suitable plants—mainly corn in the Midwest—on which to deposit their eggs. During their life span of about 12 days, they lay an average of about 1,000 eggs each, but some may lay as few as 400 and others as many as 3,000.

In the eastern United States, corn earworm pupae seldom survive the winter north of a line drawn from central Virginia to St. Louis, Missouri. Nevertheless, in summer caterpillars of this species can be abundant as far north as southern Canada. They are the progeny of moths that migrated from the south. Rick Weinzierl of the Department of Crop Sciences at the University of Illinois in Urbana-Champaign told me that the moths which lay eggs in central Illinois in June originated in Louisiana and eastern Texas.

The corn earworm, according to Robert L. Metcalf and Robert A. Metcalf,[3] may well be the most damaging pest of corn in the United States. From 70 to 98 percent of the ears of field corn are infested and as many as 7 percent of the kernels on an ear may be eaten and lost to harvest. On sweet corn, even more kernels may be eaten and lost to the harvest. (Field corn, harvested when the kernels are mature and hard, is mainly fed to livestock, although some is made into cornflakes and cornmeal. Sweet corn, harvested when the kernels are immature and soft and tender, is meant to be eaten by people.) According to the Metcalfs, each year American farmers grow the equivalent of two million acres of field corn just to feed corn earworms. There is no economically practical way to control this insect in field corn, but it may be economical to control it in sweet corn, which has a much higher value, by injecting an insecticide dissolved in mineral oil into the tip of each ear, or by carefully hand-spraying the silks with an insecticide. Spraying must be thorough and frequent because corn earworm eggs do not take long to hatch, and once the caterpillars get into the ear, a short trip along a silk, the husk protects them from insecticidal sprays.

Even as I write, entomologists and ecologists are rapidly expanding our understanding of a remarkable chemical route by which parasites are led to corn earworms or their other herbivorous hosts. There is now no doubt that some plants—and probably many more—send forth a "cry for help" when they are under attack by plant-feeding insects. Among these plants are corn, cotton, lima bean, apple, cassava, celery, cucumber, soybean, and cabbage. When injured by insects, these plants emit scents that attract and bring to the rescue parasites of their attackers. As long ago as 1955, the Canadian entomologist L. G. Monteith[4] had shown that parasitic flies are attracted to plants that have been damaged by their hosts, but it was not until the 1990s that Ted Turlings and coresearchers[5] discovered a new and amazing aspect of this line of communication between plants and parasitic insects.

A plant does not send out a call for parasites unless it is actually under attack by herbivorous insects. Plants that have been chewed on by armyworm caterpillars or their close relatives, corn earworms, release volatile chemicals, terpenoids, that attract certain wasps that are parasites of these caterpillars. The plants release these attactants, "alarm scents," only if they bear wounds that have been contaminated with oral secretions of a caterpillar. Wounding alone is not enough. Plants that were artificially damaged with a razor blade did not release the terpenoids and did not attract parasites. But when the wounds on artificially damaged plants were smeared with regurgitant from armyworms, the plants did release the volatile chemicals and attracted parasites. Uninjured plants smeared with caterpillar regurgitant did not release the attractive volatiles and did not attract parasites. Clearly, both wounding and contamination with oral secretions, as occur when an insect chews on a plant, are required to trigger the release of the chemicals that attract the parasites. The whole plant, not only leaves that have been injured, releases these attractants. Thus the plants communicate a specific message to the parasites: "Come to me and you will find hosts for your offspring."

In an engrossing article in the *National Geographic*, Gary McCracken and John Westbrook[6] tell how they used Doppler radar and hot-air balloons to track Mexican free-tailed bats and their prey. The bats migrate from Mexico to the vicinity of San Antonio, Texas, where they spend the nights chasing insects through the air and the days roosting in caves. These are essentially nursery colonies. Almost all of the bats are females and in June each gives birth to one hungry pup that is nursed by its mother, who must now eat for two. In June and July the bats feed almost exclusively on corn earworm moths and the very closely related tobacco budworm moths, which, despite their name, feed on corn and many other plants, as do their corn earworm cousins.

Early in June billions of these moths migrate from cornfields in the lower Rio Grande Valley to an agricultural area 250 miles to the north in south central Texas near San Antonio. There they lay eggs; the caterpillars do great damage to corn; and when they become adult moths, they and corn earworms from other places fly north and produce one or more generations on crops in the central United States and as far north as southern Canada.

Doppler radar showed McCracken and Westbrook that bats looking for insects fly as high as 10,000 feet. Observations made at night from a hot-air balloon that ascended to 5,000 feet convinced them that, at least at that altitude, Mexican free-tails were feeding on moths. Later, a DNA analysis of insect remains in bat feces proved that the bats had been eating mostly corn earworm and tobacco budworm moths, which, together with a few other moths, constituted 90 percent of their diet during the peak of the moth migration.

The caves in south central Texas harbor about 100 million bats. To feed herself and nurse her pup, a female bat must eat as much as 70 percent of her half-ounce body weight each night. Simple arithmetic reveals that these 100 million bats eat an incredible 1,000 tons—2 million pounds—of insects, mostly moths, in one night. Just from an economic perspective, the Mexican free-tailed bats are well worth protecting. Without them, the destructive caterpillar progeny of these moths would be distressingly more numerous.[7]

This bat and other nocturnal insect-eating bats use echolocation (essentially sonar) to locate moths and other flying prey. Bats emit intense sounds, too high-pitched for humans to hear, and sense a flying insect by listening for echoes that bounce back from its body. Moths have, however, responded to the deadly threat from bats. Most moths, including corn earworms, have evolved ears that can hear the cries of hunting bats, a forewarning that prompts them to take evasive actions that sometimes saves them. Jayne Yack and James Fullard[8] of Carleton University in Ottawa proposed that some nocturnal moths escaped from bats by evolving to become the day-flying butterflies, which lack ears because they have no need for them—except for one aberrant nocturnal butterfly, the exception that proves the rule, that retained its ears.

Readily available and easily reared on an artificial diet, corn earworm caterpillars are suitable and convenient laboratory animals for behavioral and physiological research on the amazing behavior known as nutrient self-selection, which has been anthropomorphically referred to as "nutritional wisdom." This is the ability of an animal to select for itself a balanced diet from an array, a "cafeteria," of several different foods or nutrients. "Nutritional wisdom" seems to be ubiquitous among animals, known from mollusks, fish, hermit crabs, birds, mammals including humans, and even a protozoan that eats other protozoa. But when I did my research on this behavior, there had, for some time, been only one demonstration of self-selection by an insect, the larvae of a flour beetle, and much remained to be learned about self-selection's behavioral and physiological basis.

It all began when Anoop Bhattacharya and I[9] made the serendipitous discovery that flour beetle larvae can select for themselves a balanced and nutritionally favorable diet from a mixture of large particles of the three parts of a wheat kernel: bran, the husk; germ, the embryo; and endosperm, the starchy matter that constitutes most of the kernel. The initial discovery was a fluke, a by-product of an experiment examining another matter. But, excited by our discovery, the first known example of self-selection by an insect, we set out to learn more about this insect's "nutritional wisdom."

Larvae presented with equal amounts of bran, germ, and endosperm chose an intake consisting of a negligible quantity of bran, about 17 percent endosperm, and over 81 percent germ. They fared better than did larvae given only one of these fractions of the kernel. Many larvae that had access to only bran or endosperm died, about half of those on endosperm and a third of those on bran, and the growth rate and weight gain of the survivors were abnormally low. Larvae fed only germ did not do quite as well as those that ate a self-selected mixture of germ and endosperm. They gained significantly less weight and were slightly less likely to survive.

In this study Bhattacharya and I had shown that flour beetles are able to compose an optimal diet by selecting from a mixture of foods, but, interesting as this is, it left some important questions unanswered. First, are other insects capable of this behavior? In view of the ubiquity of self-selection among other animals, we thought that they are, but science demands more than an educated guess. We needed to demonstrate self-selection by insects other than beetles. Second, what behavioral and physiological mechanisms are the basis of self-selection by insects and other animals? We have just begun to answer this latter question, proposing a plausible hypothesis that has been supported by our own experiments as well as research done by others.

The chemical composition of a wheat kernel can be analyzed but it cannot be changed. To continue this research we needed to use an artificial diet whose nutrient composition could be changed at will. This was possible with an artificial diet for the corn earworm devised by Erma Vanderzant[10] during her studies of insect nutrition. She formulated the diet by putting together separate and known quantities of nutrients, among them protein, sugar, corn oil, cholesterol, vitamins, minerals, and cellulose powder for roughage—all embedded in a gel of agar in water. (Agar, which contains no nutrients, is a gelatin extracted from seaweeds.) With this diet it was possible to leave out or alter the quantity of any of the nutrients.

With the help of several graduate students, Stanley Friedman and I[11] worked on nutrient self-selection by corn earworms for several years. Our first experiment was a great success, showing that corn earworm caterpillars selected a favorable balance of protein and sugar when offered a choice of two artificial diets that were identical and nutritionally complete except that one lacked protein and the other lacked sugar. In our experiment each one of a number of caterpillars was put in a small covered glass dish with same-size disk-shaped pieces of the two diets placed on opposite sides of the dish. We were delighted when we saw that the caterpillars moved back and forth between the two diets, feeding on the protein diet for almost four hours and on the sugar diet for about three-quarters of an hour, a ratio of about 81 to 19. When we weighed the diets, we found that the caterpillars had selected, with little variation among them, a protein to sugar ratio of about 80 to 20. We showed that this ratio is superior to others by comparing the growth and survival of caterpillars raised on diets that contained protein and sugar in the 80 to 20 or several other ratios.

Earworms in a similar experiment self-selected to obtain the necessary vitamins and the required lipid, corn oil in this case. Offered two diets, both complete except that one lacked vitamins and the other corn oil, the caterpillars always fed from both diets, usually eating large amounts of both. On the other hand, earworms offered two complete diets that contained both vitamins and corn oil fed only or mostly on just one of them.

In a similar experiment brown-banded cockroaches, common household pests, selected a protein to sugar ratio of 16 to 84, nearly the opposite of the ratio chosen by corn earworms. This remarkable difference reflects an extreme difference in the lifestyles and growth rates of these two insects. Corn earworms grow rapidly, completing the larval stage in only 17 days. They are also extremely sedentary. An earworm is likely to spend its entire larval life in one ear of corn, surrounded by abundant food and protected from its enemies. The cockroaches, on the other hand, grow very slowly, requiring about 256 days to complete the nymphal stage. Furthermore, the cockroaches are by no means sedentary. They have to be very active because they must search for shelter and scattered food resources and flee from predators or humans. The earworm requires a large daily intake of protein to support its rapid growth, but little sugar to provide energy for body maintenance and its minimal activity. On the other hand, the hyperactive cockroach requires only a small daily intake of protein, because it grows very slowly, but a very large intake of sugar to provide energy to fuel its frequent and almost frenzied activity.

What are the behavioral and physiological mechanisms that mediate self-selection? An early hypothesis maintained that animals have an instinctive nutritional wisdom. At first glance this sounds reasonable, and it's also a catchy phrase. However, if taken literally, as I have not, it implies an innate capacity to know that a particular nutrient is lacking or in short supply, to search for it, and then to recognize it when it is found. This hypothesis is difficult to take seriously, because the ability to identify each one of the over thirty vitamins, minerals, and other nutrients required by all animals would have to be hard-wired in the brain. The source of the "wisdom" must lie elsewhere. Nobel laureate Konrad Lorenz,[12] whose expertise was animal behavior, proposed a more reasonable behavioral hypothesis, that an animal tentatively eats just a little of each substance offered to it, notes "how it feels afterwards," and then feeds accordingly.

This did not quite fit with our observations, however. Corn earworm caterpillars did not eat just a little bit of either diet to find out "how it feels afterwards." They fed from the sugar and protein diets for a long time before they "decided" to switch from one to the other. We refined Lorenz's idea, and named it the "malaise hypothesis." The gist of our hypothesis is that feeding on a food lacking or in short supply of a required nutrient eventually results in a metabolic upset, which the animal perceives as a malaise, stimulating it to stop feeding and begin an exploration that may bring it to some other food which may contain the missing nutrient or have a larger concentration of it. This could, of course, explain the behavior, switching back and forth between diets, of the earworms in our experiments. Studies by others support the malaise hypothesis, most recently Robert Thacker's[13] research with land hermit crabs.

Perhaps the malaise hypothesis explains the craving for unusual foods experienced by many pregnant women. Some crave a food that they seldom or never eat under other circumstances, perhaps pickled okra or lychees. Other women, particularly in sub-Saharan Africa, eat certain types of clay or even plaster from a wall. It may be that the demands of the growing fetus deplete the mother's body of some required nutrient, causing a feeling of malaise. She knows that she requires something not supplied by her current diet, but she does not know what it is. Therefore, she "explores" by eating foods not presently in her diet, just as our earworm caterpillars begin to wander when they are sated with either sugar or protein and require one or the other of them.

But what sort of metabolic upset can cause an animal to experience a feeling of malaise? There are surely many answers to this question. Our research revealed one of them. If corn earworm caterpillars are fed nothing but an artificial diet that contains a significant excess of sugar, they digest and assimilate more of what they eat, but the proportion of the assimilated food that is devoted to the growth of the body declines markedly. The most reasonable explanation for this decline is that the earworms cannot prevent the excess sugar from being assimilated, and because it has passed through the wall of the intestines into the blood, cannot excrete it with the feces. Faced with more sugar than they can use, they must disrupt their normal metabolism by catabolizing ("burning off") the excess sugar so that it can be eliminated from the body as water and carbon dioxide.

Can insects other than flour beetles and corn earworms improve the nutritional value of their diets by feeding selectively on different parts of the same plant or on several different species of plants? Most likely all of them can and do, but only a few have been scientifically shown to do so. Several researchers found that grasshoppers routinely feed on more than one species of plant, but only a few of these researchers determined whether or not the grasshoppers benefited by feeding on several of them rather than on just some of them. But in Tohko Kaufmann's[14] exemplary experiments, caged nymphal grasshoppers with access to three species of grass gained more weight and were more than five times as likely to survive to the adult stage than grasshoppers with access to just some of these grasses. At least some predators are also capable of self-selection, but little is known about this. However, most spiders, eight-legged relatives of the insects, must, as reported by Susan Reichert and Joel Harp,[15] experts on spiders, eat more than one prey species in order to survive.

The corn earworm is one of the most troublesome pests of field crops in North America, but it is only one among many. In the United States and Canada alone there are well over 250 other species of insects that, to varying degrees, injure our field and vegetable crops, our shade trees, and our garden flowers. Quite a few of these pests were unintentionally introduced into North America from abroad. Seventeen of the most destructive of them were introduced, during a period of about 100 years, from 1889 to 1986. Among them is the horrendously pestiferous gypsy moth, which came to us from Europe and is the leading character of our next chapter.

12
INVADERS FROM ABROAD
Gypsy Moth

On a day in May, some time in the 1950s, I ventured a little way into a Massachusetts woodland that was heavily infested with caterpillars of the infamous gypsy moth (*Lymantria dispar*). The leaves of the tall trees, silhouetted against the sky, were ragged and had already suffered significant damage, although most of the caterpillars were barely half-grown and had as yet eaten only a small fraction of what they would ultimately eat. It was obviously and unpleasantly apparent that hundreds of thousands, if not millions, of the hairy larvae were feeding voraciously up in the trees. Fecal pellets rained down like a summer shower; the chewing of the caterpillars' jaws made an audible and pervasive rustling sound; like parachutists, caterpillars by the hundreds dropped from the trees on strands of silk, many landing on me and some going down the back of my neck. When I revisited this woodland in June, the caterpillars were gone, having molted to the pupal stage, but they had all but completely denuded the trees, leaving only tiny green stubs of the leaves, mainly stems and sometimes a part of the midrib. From a distance, the bare, gray trees—showing just a light flush of green, the leaf stubs—looked like a woodland in early spring when tiny, new leaves are just beginning to unfurl.

The gypsy moth is one of the very few introduced pest insects whose exact "port and date of entry" to North America is known. In 1916 A. F. Burgess of the US Department of Agriculture Bureau of Entomology wrote:

> Many years ago a circumstance occurred at Medford, Mass., which was destined to cause enormous expense and trouble in that community and throughout the neighboring States. About 1869, Prof. Leopold Trouvelot, a French naturalist who was a resident of Medford [in a house now known as 27 Myrtle Street], introduced a few egg clusters of the gipsy [*sic*] moth for the purpose of conducting experiments on silk culture. During the course of the experiments some of the caterpillars escaped. Realizing that the insect was a serious pest in Europe, he made a careful search on the trees and in the woodland nearby for the purpose of destroying any that could be found. He also notified the Department of Agriculture at Washington. None of the insects which had escaped could be found, but as no injury resulted during the next few years, it was thought that the matter was not of great importance.
>
> About 20 years later the neighborhood was invaded by the swarms of caterpillars which were supposed by most of the residents to be a native species that had become unusually abundant. A study of the matter developed the fact that the insect which was defoliating the trees was the notorious gipsy moth of Europe and that it had become firmly established in the locality in which it had originally escaped and throughout the immediate surroundings.[1]

In their account of the first troublesome infestation of gypsy moths in North America, which occurred in Medford in 1889, Edward Forbush, who directed the first attempts to eradicate this invader from Europe, and Charles Fernald, an entomologist at the Massachusetts Agricultural College in Amherst (now the University of Massachusetts) quoted interviews with residents of Medford that make graphically obvious the horror of an urban infestation of gypsy moths. Mrs. William Belcher said:

> My sister cried out one day, "They [the caterpillars] are marching up the street." I went to the front door, and sure enough, the street was black with them, coming across from my neighbor's, Mrs. Clifford's and heading straight for our yard. They had stripped her trees, but our trees at that time were only partially eaten.

Forbush and Fernald summarized:

> The number of caterpillars that swarmed over certain sections of the town during the latter part of June and most of July, 1889, is almost beyond belief. Prominent citizens have testified that the "worms" were so numerous that one could slide on the crushed bodies on the sidewalks; and that they crowded each other off the trees and gathered in masses on the ground, fences and houses, entering windows, destroying flowering plants in the houses, and even appearing in the chambers at night. The huge, hairy, full-grown caterpillars were constantly dropping upon people on the sidewalks beneath the trees, while the smaller larvae, hanging by invisible threads, were swept into the eyes and upon the faces and necks of passers. The myriads that were crushed under foot on the sidewalks of the village gave the streets a filthy and unclean appearance. Ladies passing along certain streets could hardly avoid having their clothing soiled, and were obliged to shake the caterpillars from their skirts. Clothes hanging upon the line were stained by the larvae, which dropped or blew upon them from trees or buildings. In the warm, still summer nights a sickening odor arose from the masses of caterpillars and pupae in the woods and orchards, and a constant shower of excrement fell from the trees. The presence of this horde of gypsy-moth larvae had become a serious nuisance, and was fast assuming the aspect of a plague.[2]

Since its North American debut, the gypsy moth has spread widely. It now occurs in Ontario, Quebec, and much of the US mainland: as far north as Maine; south to Maryland, Virginia, and North Carolina; westward to Minnesota and Texas; and isolated infestations have been found in California, Oregon, and Washington. Spot infestations found in other states have been eradicated, but it is a losing battle. Eventually the gypsy moth will infest suitable forests, including urban forests of shade trees, everywhere in the United States and southern Canada. It has become a terribly destructive pest of fruit, shade, and ornamental trees, and D. Barry Lyons and Andrew Liebhold[3] said that it "is the most significant pest of hardwood forests in North America." The caterpillars are known to feed on well over 500 species of shrubs and trees, including both hardwoods and pines. They do not find all of these plants to be equally palatable, as pointed out by W. E. Britton[4] of the Connecticut Agricultural Experiment Station in New Haven. They seem to be particularly fond of certain plants, among them fruit trees, roses, oaks, elms, willows, poplars, aspens, and birches. Although the caterpillars cannot survive on some trees when they are young, when they are older they readily feed on them and thrive. "For instance," explains Britton, "though the caterpillars in their later stages will feed upon pine, and have killed large areas of pine mixed with hard wood in Massachusetts, they must have some deciduous foliage for food in their first and second larval stages."

The severity of gypsy moth infestations varies from year to year and from area to area, according to Michael Gerardi and James Grimm.[5] A locality may be only lightly infested and suffer insignificant damage for one or more years, but in another year it may be heavily infested and severely defoliated. The total area of forest severely defoliated fluctuates greatly from year to year. In three successive years in New England, for example, it was 1.5 million acres, 491,000 acres, and 52,000 acres.

Defoliation can kill trees. The larch is exceptional in that it can withstand repeated defoliation for 10 years or more. Pines die after only one defoliation. Deciduous trees may, if healthy and vigorous, survive a few successive years of defoliation, in some species only two or three. Partial defoliation seldom kills a tree, but weakens it, thereby stunting its growth and lowering the commercial productivity of the forest.

The defoliation of forest trees also has far-reaching ecological effects. More nestling birds die because they are not shaded from the heat of the sun. Consequently, trees lose some protection against their insect enemies as the population of insectivorous birds declines after a gypsy moth infestation. The increase in soil temperatures caused by the loss of shade affects, in some cases harmfully and in others beneficially, the understory of seedling trees, shrubs, and herbaceous plants as well as other organisms that live within or on the soil. The heat of the soil surface causes snakes to move to lower elevations or more shaded places. Creatures such as wild turkeys, squirrels, raccoons, opossums, and bears suffer from a shortage of food as acorn production declines because oaks have been defoliated.

Gypsy moths have only one generation per year. They spend nine months of their lives, from August to the following May, in the egg stage. Oval, inch-long masses of about 400 diapausing eggs—but sometimes as many as 1,000—covered with light yellowish hairs from the underside of the mother's body, are attached to tree trunks, fences, buildings, vehicles, or anything else that offers support. In southern New England, caterpillars begin to hatch at the end of April. Then only about a tenth of an inch long, they rest on the egg mass for a few days before crawling to a leaf to feed voraciously. Full-grown caterpillars, as much as three inches long, crawl about to find a protected place, often on the trunk of a tree, where they spin a scanty silken cocoon in which they molt to the inactive pupal stage. After from 10 to 14 days, the pupa molts its skin and the adult emerges. As described by Britton, "The male moth is generally brown in color and flies about even in the day time. The female is nearly white, with rather inconspicuous cross markings of brown and black, and is larger than the male. The female has a heavy body and does not fly but rests on the trunk or branches of trees, sides of buildings and walls, and in such places she lays her eggs. . . . The adults take no food and live only a short time."[6]

Females attract males by releasing a chemical signal—a volatile,

A female gypsy moth and her recently laid egg mass on the bark of an oak tree.

airborne pheromone—hastening its evaporation from their body by fanning their wings. Males perceive infinitesimal concentrations of the pheromone, known as gyptol, and by flying upwind into a plume of gyptol, can locate a female from a considerable distance—as far as

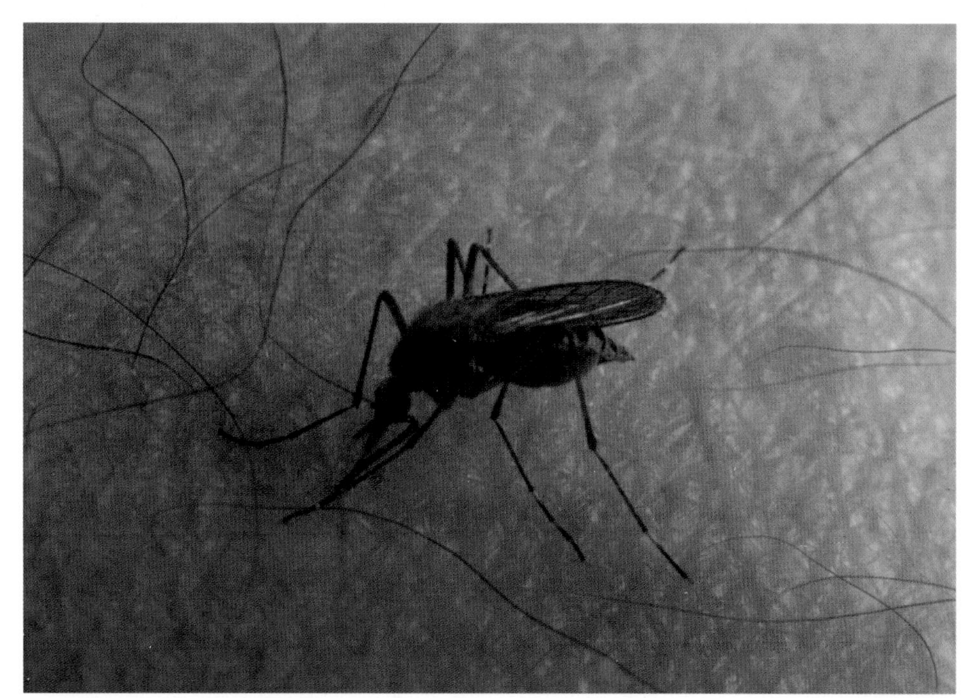

A female mosquito filled with blood sucked from the photographer's arm. *(Photograph by James Sternburg)*

A house fly perched on the wall of a barn. *(Photograph by Philip Nixon)*

Adult northern corn rootworms feeding on a thistle blossom. *(Photograph by James Sternburg)*

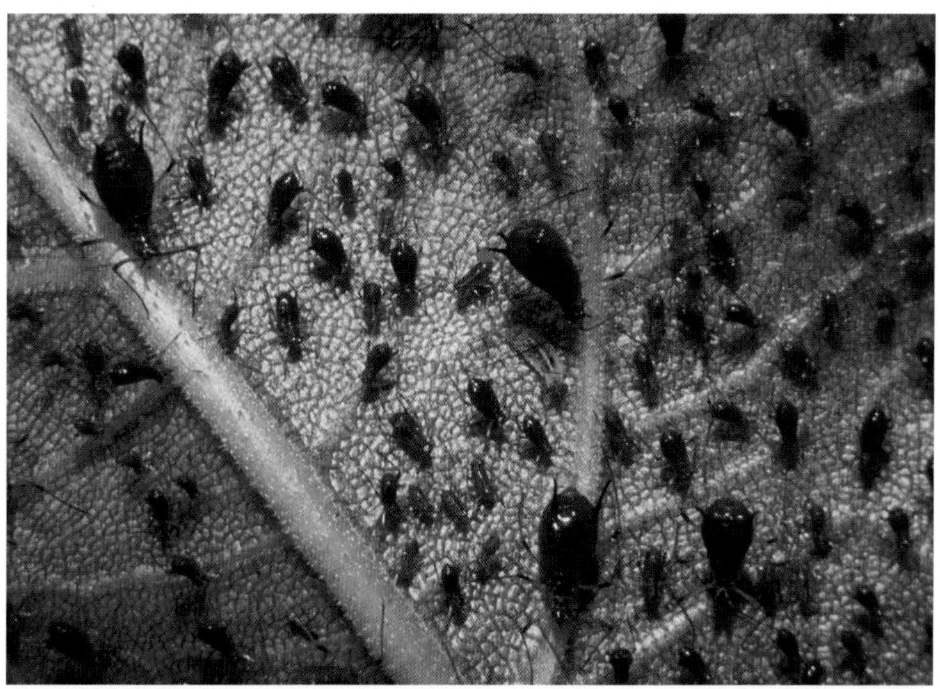

A colony of aphids; in the center is a predator eating a young aphid. *(Photograph by James Sternburg)*

Bagworm cases hanging from a juniper twig. The pupal skin of a male protrudes from one. *(Photograph by Philip Nixon)*

A small swallowtail larva that looks like a bird dropping. *(Photograph by James Sternburg)*

A large black swallowtail larva that mimics a monarch caterpillar. *(Photograph by James Sternburg)*

Camouflaged black swallowtail pupae. *(Photograph by James Sternburg)*

A black swallowtail female resting on a fern. *(Photograph by James Sternburg)*

The nonmimetic form of the female tiger swallowtail. *(Photograph by James Sternburg)*

The form of the female tiger swallowtail that mimics the pipevine swallowtail. *(Photograph by James Sternburg)*

Cabbage white butterfly sipping nectar from an aster. *(Photograph by James Sternburg)*

A corn earworm caterpillar munching the kernels of an ear of sweet corn. *(Photograph by Philip Nixon)*

A corn earworm moth with its tongue poked into a blossom. *(Photograph by James Sternburg)*

A hairy gypsy moth caterpillar lying on a glass plate with leaves that it has fed on. *(Photograph by Philip Nixon)*

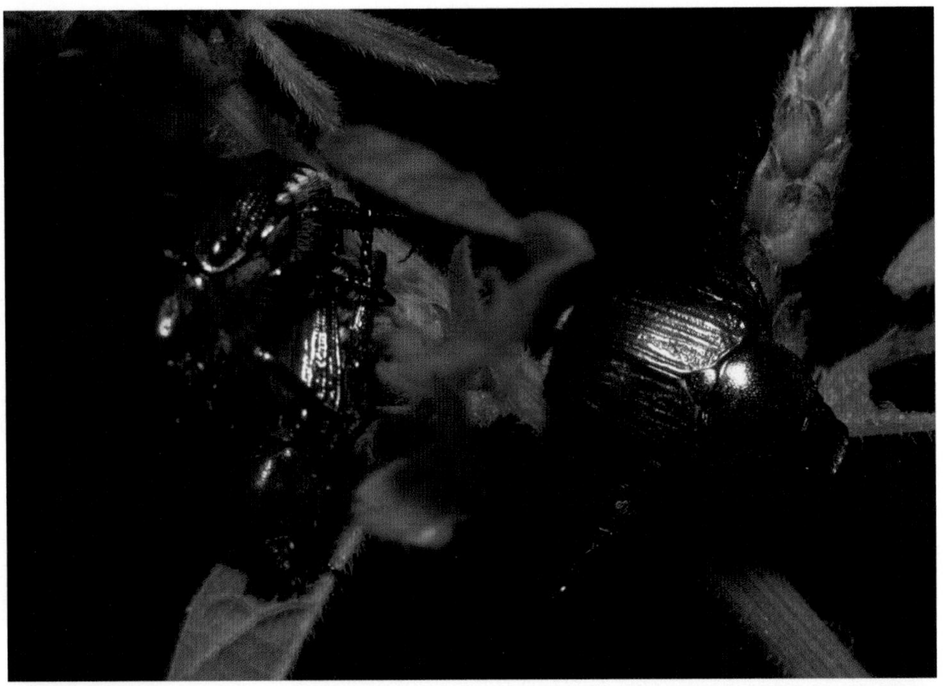

A mating pair of Japanese beetles with a disappointed suitor nearby. *(Photograph by Philip Nixon)*

seven miles, according to Gerardi and Grimm.[7] But much of this record-breaking male's route to the female was surely random wandering rather than the following of a seven-mile-long pheromone odor trail. Gyptol was chemically characterized and synthesized in 1969.[8] As you will see below, synthetic gyptol, known as disparlure, is an important tool in programs to control gypsy moths.

Chemical signals—pheromones—used not only by insects but also by mammals such as dogs and even humans, are perceived by and affect the behavior or physiology of another individual of the same species. Pheromones have many different functions. When attacked by a predator, aphids emit a warning pheromone that puts their nearby siblings on alert. The queens in colonies of social insects such as bees and ants produce a pheromone that prevents workers of the colony from reproducing by blocking their sexual development. A worker ant that finds a good source of food marks her way back to the nest by daubing the ground with a trail pheromone. Other workers follow this trail back to the food and—as long as food remains—reinforce the pheromone trail on their way back to the nest. This is why we see files of ants—sometimes in our kitchens—follow precisely the same path as they go back and forth from their nest, which may be hidden in a wall, to a bit of spilled sugar or some other food. You can disorient and drive the ants to a frenzy by wiping away a section of their chemical trail with a cloth wet with alcohol.

How did gypsy moths spread over the vast area they now occupy when the flightless females lay their eggs no more than a few feet from where they themselves were hatched? All organisms must have some way of dispersing—at the very least for short distances. The offspring of a firmly rooted burdock plant (its seeds) may be carried for miles in a spiny seedpod tangled in the fur of an animal. The quintessentially sessile barnacle—a very unusual arthropod related to the crabs—is immovably fixed to a rock but begins life as a free-swimming, wandering larva. Some insects with flightless mothers, including the bagworm and the gypsy moth, disperse as tiny larvae, floating kitelike on the breeze suspended from a long thread of silk. Shortly after hatching, gypsy moth caterpillars drop down on a thin strand of silk to be carried away by the wind. Most do not travel very far, but a few may land 20 miles or more from where they began their journey.

We humans have inadvertently transported gypsy moth eggs for hundreds or even thousands of miles. As long ago as 1916, Burgess noted that egg masses, which the females place on any convenient surface, may be transported from place to place on vehicles, and observed that "Since the gipsy-moth [*sic*] [control] campaign first began, an unprecedented development in means of rapid transportation has taken place. At first and for several years motor vehicles were practically unknown, but for the last few years the increase in this mode of transportation has been enormous."[9] In 1909 Henry Ford introduced the first reasonably priced automobile, the famous Model T; 15 million of them were manufactured during the next 19 years. At that time there were few paved roads and long-distance travel by automobile was not common. But it was not long before thousands of miles of paved roads were built, the number of motor vehicles soared, and people made automobile trips of hundreds or even thousands of miles. Today commerce and leisure travel carry gypsy moth egg masses just about everywhere in North America. Consider the following scenario: In July of 1992 a gypsy moth lays her egg mass on the underside of the fender of a camper parked in Rhode Island. The following spring a vacationing family drives the camper to the shores of a lake in Wisconsin and settles in for a week of fishing and swimming. Unknown to them, the gypsy moth eggs, about 400 of them, hatch and the caterpillars move to nearby trees. Only 12 survive to become egg-laying females, but they will lay 5,000 or more eggs that will hatch the following spring.

The people of Medford and nearby towns invaded by the gypsy moth soon decided that enough was enough. Not only was this insect appallingly destructive and an aggravating nuisance, but, as Burgess reported, "In some sections it was impossible to rent property on account of the abundance of the caterpillars, and real estate values declined rapidly."[10] The people demanded action. The city of Medford was the first to allot funds, and in 1890, the Massachusetts state legislature also appropriated money to pay for the fight against the gypsy moth.

In the nineteenth century there were no "miracle" insecticides, and other methods had to be used to kill the gypsy moths—methods that are labor-intensive but were and are still highly effective. The egg masses are obviously the most vulnerable stage of the insect's life cycle. They are accessible for the nine months of the year when they are the only living representatives of their species. Furthermore, they are conspicuous because they are large, almost white, and usually found easily because they are in plain sight on the trunk of a tree, a fence, or some other support. Consequently, gypsy moth populations can be controlled or perhaps even eradicated by destroying their eggs.

At first egg masses were scraped off with a blade, but that was unsatisfactory because some eggs broke loose from the mass, fell to the ground, and survived to hatch in the spring. Of the several other methods that were tried, the simplest and most efficient was to daub the egg masses with creosote, which quickly penetrated the mass and killed the eggs. This was not always easy to do because some masses were too high to reach by hand and others were hidden, perhaps under a fallen log or a porch.

Some were impossible to reach because they were hidden in interstices of the walls of piled rocks so typical of New England. But that problem was solved by the invention of the "cyclone burner," an early version of the flamethrowers so cruelly but effectively used in combat during World War II. The cyclone burner consisted of a large, closed vat of oil kept under pressure by a man wielding a hand pump, and a long hose that ended with a metal tube with a "cyclone nozzle" at its end. The oil that blasted from the nozzle was ignited by a second man who directed the hot flame into the cracks and crevices of the stone wall, incinerating any egg masses that were there.

Caterpillars were controlled in two ways, by keeping them from climbing up the trunks of trees and by poisoning them with an insecticidal spray of lead arsenate. Bands of sticky material encircling the trunk trapped climbing caterpillars. Paper bands coated with tar were effective, but even more effective and much easier to apply was a band of very sticky material, known as *Raupenleim*, squirted directly on the bark. The German word *Raupenleim* means "caterpillar lime," lime in the same sense as the sticky birdlime that has been painted on twigs to trap small perching birds. A similar substance, Tanglefoot, is used today in conjunction with research or control projects with many different kinds of insects. Of the few insecticides available at the time, most of them arsenic compounds, lead arsenate proved to be the least objectionable and most effective killer of gypsy moth caterpillars. When mixed with water in concentrations high enough to consistently kill caterpillars, other arsenicals burned and killed the leaves on which they were sprayed, but lead arsenate could be used at any desired strength without serious injury to the foliage.

The campaign against the gypsy moth was very effective. "This work," Burgess wrote, "reduced the infestations to such an extent that many citizens who . . . had been seriously annoyed by the pest, or had suffered severe loss from it, came to the conclusion that because it was seldom seen the work was unnecessary and no harm would result if measures for its control were discontinued."[11] This conclusion was based on the witless assumption that there were no caterpillars because none were obvious to the eye. In the fall of 1899 enough political pressure was "brought to bear . . . to cause the discontinuance of state appropriations for the control of these insects." In 1900, when the organized control program ended, the infestation covered only 359 square miles in the vicinity of Boston. With relatively little expense the gypsy moth could have been contained in this rather small area, and with perseverance—and perhaps a bit of good luck—it might even have been eradicated, freeing all of North America from this noxious invader. Terminating the control program was a costly mistake. By 1905 the gypsy moth had spread widely and was again such a nuisance that the Massachusetts legislature was forced to appropriate funds to resume the campaign against it. But by that time the infestation covered well over 2,000 square miles, making containment very difficult and making eradication all but impossible. Gypsy moth control is not cheap. Just from 1906 to 1934, it cost the federal government alone over $40 million (equivalent to about 400 million current dollars), and from 1925 to 1936 almost four million acres of New England forest were significantly or totally defoliated.

Soon there were gypsy moth infestations in several of the New England states, in 1913 prompting the federal government to impose a quarantine that regulated the shipment of materials from infested to uninfested areas. In 1923 a barrier zone to prevent the gypsy moth from moving westward from New England was created east of the Hudson River. Extending from the Canadian border south to Long Island Sound in New York, it was 250 miles long, from 25 to 30 miles wide, and had an area of about 8,000 square miles. Every effort was made to eradicate spot infestations that appeared in the barrier zone. Although small infestations, which were quickly eradicated, appeared as far west as Ohio, the barrier zone confined the gypsy moth to New England until 1952, when a small area in New York west of the barrier became permanently infested. By 1973 New Jersey and most of New York and Pennsylvania, as well as a small area in Michigan, were also infested. Since then, the gypsy moth, as already noted, has spread to other states and Canadian provinces and continues to spread. The fight against this noxious insect continues, using the synthetic insecticide carbaryl, which does not burn foliage and is much less toxic to vertebrates than lead arsenate, and also using a naturally occurring insecticide, B_t, which is contained in the spores of a bacterium, *Bacillus thuringiensis*. B_t does not injure foliage, is totally harmless to warm-blooded animals, and because the strain of B_t used is toxic only to larvae of moths and butterflies, it spares parasitic and predaceous wasps and flies—some of which attack gypsy moths. Disparlure, the synthetic sex pheromone, has proved to be a valuable tool in the struggle against the gypsy moth. It has been applied to large areas to confuse mate-seeking males; traps baited with it have been used to reduce populations by killing males; but it has probably been most useful as a bait for traps used to detect new infestations or to monitor the effectiveness of control programs.

Beginning in 1905 there has been a continuing effort to import parasites, predators, and pathogens of the gypsy moth from its homeland in Eurasia in the hope that once they became "naturalized citizens" of North America, they would reduce gypsy moth populations, a tactic known as biological control. The entomologists waging war on the gypsy moth were inspired by the fabulous success of biological control, only 15 years before, in saving the citrus orchards of California from certain destruction by a foreign insect, the cottony cushion scale. This was, as you will see in chapter 19, the first and probably the most successful biological control of an insect.

A complex of exotic insects that prey on or parasitize the eggs, larvae, or pupae of the gypsy moth have been successfully established in the United States and Canada. But in 1935 Britton observed that "the combined efforts of both native and introduced parasites and natural enemies have not been sufficient to prevent the gypsy moth from defoliating trees over large areas,"[12] and as recently as 1994, Marjorie Hoy,[13] a University of Florida entomologist, wrote that biological control has been only partially successful in reducing gypsy moth populations. In Eurasia, observed Gerardi and Grimm,[14] this insect is attacked by about 100 known insect parasites and predators, but in North America it has far fewer natural enemies, either native or introduced. Nevertheless, the gypsy moth is as destructive in Eurasia as in North America, suggesting that it might not be more than partially controlled even if all 100 of its Eurasian natural enemies were introduced and became established in North America. From 1972 to 1974, despite biological control and the use of insecticides, the gypsy moth defoliated nearly four million acres of forest in New England, New York, New Jersey, and Pennsylvania.

From 1906 to 1986 a parasitic Eurasian fly of the family Tachinidae, *Compsilura concinnata*, was repeatedly released in North America in the hope that it would control the gypsy moth. In the larval, or maggot, stage, it lives within the bodies of the caterpillars or pupae, killing them. Unfortunately, *Compsilura* is not a specialist that attacks only the gypsy moth. Experts on the family Tachinidae, Curtis Sabrosky and Paul Arnaud Jr.,[15] report that it is a generalist that parasitizes caterpillars of at least 200 other species, only a few of which are pests. Because it has three generations per year, coinciding with gypsy moths only during their one generation in spring, it requires alternate hosts at other times. But it also parasitizes other caterpillars even when gypsy moths are available.

Compsilura parasitizes the larvae of native moths, but with just a few exceptions, little is known about its effect on them and the ensuing effects on their ecosystems. But bear in mind that there may be a significant ripple effect. For example, if *Compsilura* sufficiently reduced the population of a caterpillar that limits population increases of its host plant, that plant could then outcompete other plants, perhaps significantly altering the composition of the ecosystem's plant community, and in turn affect the many organisms that directly or indirectly depend upon these plants.

Because of *Compsilura*, populations of at least two of our native silk moths, the beautiful cecropia (with a wingspan of up to six inches, our largest moth) and the somewhat smaller but also beautiful promethea are in steep decline. Not long ago both of these moths were so common everywhere in the Northeast—but never abundant enough to be destructive—that in winter it was possible to collect their cocoons by the basketful. When Jim Sternburg and I were doing research on these two moths in the 1960s and 1970s, they were abundant. In the cities of Champaign and Urbana, Illinois, we collected 721 and 1,073 of their cocoons in the winters of 1967–1968 and 1968–1969, respectively. Beginning in the early 1980s, they became so scarce that it is now almost impossible to find one. Promethea was even more abundant. In just one day we collected over 1,000 of their cocoons near Medaryville, Indiana, with no effect on the following year's population. Just last year I went back to Medaryville and found only eight cocoons, all of which contained dead larvae apparently killed by parasites.

Several hypotheses have been advanced to explain the precipitous decline of these moths, but the results of field experiments reported by George Boettner and his coauthors[16] in 2000 show that *Compsilura* is the culprit. Of 500 cecropia larvae they placed in trees in a Massachusetts woodland, none survived to molt to the pupal stage, over 81 percent of them having been killed by *Compsilura*. The results with promethea were similar. Of 657 larvae exposed on trees for only six days, almost 83 percent were parasitized by *Compsilura*. So far we do not know if the decimation of cecropia and promethea populations has had a ripple effect in their ecosystems. But even so, the virtual loss of these beautiful moths has had an impact on people. They were admired for their size and beauty, and many a budding naturalist was thrilled and enlightened by watching a moth emerge from a cocoon he or she had collected in the field.

The gypsy moth is neither the first nor the last pest insect to be introduced into the United States from abroad. In the year 2000 David Pimentel, a Cornell University entomologist, and his coauthors[17] reported that about 4,500 arthropods (insects and their relatives such as mites and spiders) have been introduced into the United States, about 95 percent of them unintentionally. Over 2,500 of these exotic species now live in Hawaii and over 2,000 in the continental United States. Many of the latter are insects and mites that attack crops. Pest insects are a tremendous burden on our economy. Every year they destroy about 13 percent of potential crop production in the United States at a cost of almost $35 billion, about $14 billion of which can be ascribed to introduced insects. The next time you go through customs when reentering the United States, know that there is good reason why you are asked if you are bringing in any fruit or other plant materials. Some South American coffee berries were seized by customs from a Florida schoolteacher who wanted to show them to her students. It was a good thing they were not brought into the country. When later examined they were found to be infested by larvae of the Mediterranean fruit fly. Once again establishing this fly in Florida would have been a disaster for Florida fruit growers.

Among the more infamous stowaways from abroad is the European Hessian fly, a destructive pest of wheat, that was first found in this country on Long Island in 1779. The destructive, leaf-feeding Japanese beetle, first discovered in a nursery in New Jersey in 1916, sneaked in on the roots of nursery stock from Japan. The European corn borer, the target of the controversial genetically modified corn that contains genes of the bacterium that synthesizes B_t, was a stowaway in broomcorn imported from Italy or Hungary in 1917. Larvae of the Asian tiger mosquito, which is capable of transmitting several disease-causing viruses, came into the country as larvae in water trapped in used tires from China, Japan, and South Korea. It was first seen near Houston, Texas, in 1985. The wood-boring Asian long-horned beetle, a serious threat to urban and woodland hardwood trees, came to us in the green lumber of shipping crates from China, and was first seen in Brooklyn, New York, in 1996.

Pest insects as well as harmless ones move in both directions across the Atlantic. Insects may contaminate plants, seeds, grain, wood, or other commodities that cross the ocean, and with airplanes that cross the Atlantic in just a few hours, an insect—perhaps a pregnant female—can be carried in a passenger's luggage or even survive the crossing in the passenger compartment, or on a passenger's jacket. While the gypsy moth and many other insects came to North America from Europe, others, among them the Colorado potato beetle, the western corn rootworm, and (the subject of the next chapter) the grape phylloxera, arrived in Europe from North America. The grape phylloxera began to destroy grapevines in Europe in the 1860s. At that time the modern "miracle insecticides" did not exist. A good thing, I think, because an alternative control that was nonpolluting and did not require the repeated application of insecticides year after year was made possible by studying the natural history of the grape phylloxera, especially its association with grapevines of both the American and European species.

13 AN AMERICAN SAVES THE FRENCH WINE INDUSTRY

Grape Phylloxera

Since time immemorial, wine has been made by the people of Europe and the Near East, particularly in areas favored by the Mediterranean climate that makes wine grapes thrive. In ancient Greece and Rome, Bacchus was recognized as the god of wine. Genesis 9:20 tells us that "Noah began to be a husbandman, and planted a vineyard: and he drank of the wine and was drunken. . . ." Many centuries later Jesus turned water into wine. By the nineteenth century the French were master vintners, producing many different red and white wines from a remarkable diversity of cultivated grape varieties. In his authoritative and eminently readable *The Great Wine Blight*, George Ordish pointed out that all of the varieties of grapes cultivated in Europe are races of the one and only European species of grape, *Vitis vinifera*, that is fit for making wine.[1] But in the 1860s the French wine industry was threatened with imminent destruction by an American insect.

This aphidlike insect, the grape phylloxera (*Daktulosphaira vitifoliae*, formerly known as *Phylloxera vitifoliae*) was first noticed in France in 1863 and soon became the scourge of the wine industry. It quickly spread to other European countries and ultimately to wine-producing areas all over the world. In Europe it very seldom causes galls (tumor-like swellings) on the leaves as it does in the United States on other species of grapes; the real damage is done by a form of the phylloxera that induces the formation of galls on the roots. The roots soon rot and, as a consequence, the foliage of the vines turns yellow, the vines lose vigor, and they eventually die. Edward H. Smith,[2] emeritus professor of entomology at Cornell University, wrote that the unintentional introduction of the grape phylloxera—probably brought in on rootstocks from the United States—was an unmitigated disaster for French viticulture. Within 25 years of its appearance, the phylloxera had destroyed nearly a third of the vineyards in France, more than 2.5 million acres, and there was no reason to hope that the destruction would stop. It looked as if the French wine industry was doomed. French culture without the lubricating effect of fine wine is unthinkable. Furthermore, the loss of the wine industry would have had a disastrous effect on the economy of the country. It provided the livelihood of one-sixth of its population and produced a quarter of its agricultural income.[3]

In 1871 French authorities, unable to stop the destruction caused by the grape phylloxera, invited the great American entomologist Charles Valentine Riley—who probably knew more about this insect than anyone else—to visit France to help find a way to control the pernicious phylloxera. Riley saw the ever-worsening effects of this insect, and with French entomologists and leaders of the wine industry formulated plans for coordinating a joint American and French attack on the problem. "Riley," wrote Smith,

> returned from France . . . filled with ideas. . . . In the effort that followed, Riley had little interaction with fellow entomologists in the United States. There were able colleagues, but without organized viticulturalists clamoring for help and with no incentive to aid the French, the phylloxera took a low priority.[4]

The common aphorism "know your enemy" describes the foundation of economic entomology—the science that seeks to protect us and the things we value against destructive insects. We cannot take rational steps to control an insect unless we know its identity and understand it and its ways. Riley knew this well and was dismayed because so little was known about the grape phylloxera. There was even controversy about its scientific name and disagreement about whether it was an aphid or a scale insect. But most troublesome of all was that practically nothing was known about its behavior or its life cycle. In cooperation with French entomologists, Riley worked out its unusual life cycle.

The grape phylloxera, which feeds only on the roots or leaves of plants of the grape family, has an exceptionally complex life cycle. All species of aphids, including the phylloxeras, have complex life cycles but the grape phylloxera's is probably the most complex of all. Compounding the difficulty of unraveling this insect's life cycle is the fact that it differs according to the species of grape with which it is associated. On American grapes, of which there are 18 species, the phylloxera has both underground and above-ground phases of its life cycle and reproduces both sexually and parthenogenetically. But on the European wine grape, reproduction is only parthenogenetic and the phylloxeras live almost exclusively underground on the roots.

Riley's explication of this insect's life cycle on American grapes was succinctly summarized by John Henry Comstock[5] of Cornell University, the first of the great teachers of entomology in the United States. The complete cycle extends through two years and includes four different adult types. In the spring of the first year, fertilized eggs that had been laid on woody grape stems the previous fall hatch. The young nymphs crawl to leaves and form galls in which all of them mature to become adult, wingless, parthenogenetic females, called stem mothers, that fill their galls with eggs and then die. Their offspring, all parthenogenetic and wingless females like their mothers, give rise to several similar leaf-galling generations. In fall, offspring of the leaf-gallers crawl down the stem to the roots of the vine and diapause. In spring of the second year, they form galls on fibrous roots and mature to become wingless, parthenogenetic, root-dwelling females, the second adult type and progenitors of several similar root-infesting generations. The third adult type appears in the fall, when some of the root-dwellers produce parthenogenetic *but winged* females that leave the soil, fly to other vines, and lay eggs of two sizes on the woody stems. These eggs hatch, producing the fourth and last type, wingless but sexually reproducing females from the large eggs, and from the small eggs, the only males in the entire cycle. After mating and insemination, the female lays just one fertilized egg, placing it on a woody grape stem. It will spend the winter in diapause and produce a stem mother in spring, thereby completing the two-year life cycle.

Riley knew that even though the grape phylloxera commonly occurs on American species of grapes, it does not kill them as it does the wine grapes of France. He reasoned that the susceptibility of the European grape and the resistance of the American grapes is an example of Darwinian natural selection. Although many of his colleagues did not accept this idea—at that time the theory of evolution was still being debated by biologists—he was, of course, correct. The grape phylloxera and the American species of grapes, coexisting for millennia, have coevolved and no doubt continue to do so. In other words, they have a long history of reciprocal responses, the grapes evolving chemical and other defenses and the insects reacting by evolving ways of circumventing these defenses. Consequently, American grapes are now, to varying degrees, resistant to the phylloxera. These tiny insects can survive on them but do not become so numerous as to cause serious injury. On the other hand, the European wine grape, which did not coevolve with the phylloxera, had not become resistant to it and helplessly succumbed to its attack.

In 1873, in response to an invitation from Riley, the French entomologist Jules Planchon visited the United States to tour the vineyards of the eastern seaboard and to meet with Riley to lay plans to jointly exploit the phylloxera resistance of American grapes to control the phylloxera in France.

Growing American species of grapes in France and making wine from them was not a practical option, for Europeans disliked wines made from American grapes because of their "foxy" flavor and thought them to be unsuitable for the table. But there was another solution, grafting the susceptible French varieties on rootstocks of resistant American grapes.

Though the roots of the grafted plants did not succumb to the grape phylloxera, the vines at first did not grow well in France. The American rootstocks, suited to the acid soils of the United States, were not adapted to the chalky, alkaline soils of France. This difficulty was solved by not using pure American rootstocks, but rather rootstocks that were hybrids between a resistant American species and the French wine grape. These horticulturally superior hybrid rootstocks solved the problem and then the replanting of the French vineyards with grafted plants began.

This was a colossal job. When it began in 1880 there were in France over 11 trillion grapevines that had to be replaced. "This," as Smith put it, "required a vast new infrastructure to obtain American cuttings, induce them to produce roots, transport, graft, and finally set them in the vineyards."[6] And then the American plants had to be hybridized with European plants. Despite the immensity of the task they faced, "the viticulturists of France forged ahead with dogged determination."

By the early 1890s the replanting was well on its way to completion, and it was obvious that victory over the phylloxera was not far off. The winegrowers and the people were jubilant. In 1894, six years after he died, a monument honoring Planchon was erected in the heart of Montpellier. It shows a vineyard worker presenting a splendid bunch of grapes to Planchon and bears the inscription, "The American vine made the French vine live again and triumphed over phylloxera."

In 1873 the French government awarded Riley their grand gold medal and made him a Chevalier of the Legion of Honor. In 1892 the viticuturists of France expressed their gratitude to Riley by presenting him with a statuette by Henri Rousseau of a man and a woman carrying baskets overflowing with plump grapes, a depiction of the abundance of the harvest. The statuette can now be seen in the foyer of the headquarters of the Entomological Society of America in Landham, Maryland.

Happy with their defeat of the grape phylloxera, the French resumed producing some of the world's finest wines. But unknown to them, there was a cloud on the horizon, a resurgence of the grape phylloxera that began in Spain in 1915 and soon spread to other parts of the continent. Evolution, including the escalating arms race between insects and plants, is a continuing process that is going on right now. The phylloxeras began the next battle in the arms race. Just as a few house flies had an enzyme that by happenstance detoxified DDT, a few phylloxeras were able to establish themselves on the hybrid rootstocks onto which the French grapes had been grafted. The way they overcame the resistance of the plant is largely unknown. These resistant individuals were the parents of succeeding generations of resistant phylloxera that soon replaced the population that had been deterred by the American rootstocks. The phylloxera problem was back, but was solved by grafting wine grapes onto different rootstocks to which the phylloxera is not resistant. Understanding how the phylloxera manages to survive and even thrive on rootstocks that were once resistant to it is not possible unless we understand the physical and biochemical mechanisms of the plant's resistance. Unfortunately, very little is known about these mechanisms in grape rootstocks. But Jeffrey Grannett of the University of California, a leading expert on the grape phylloxera, told me that newly hatched phylloxera nymphs are able to burrow in the soil and find grape roots, but on some types of rootstocks

The statue awarded to Charles Valentine Riley by the grateful winegrowers of France.

can seldom so much as begin to establish themselves on the roots they find. They do not feed, do not form galls, and soon starve to death, suggesting that the roots contain biochemicals that deter phylloxeras before they can attack the roots. It could also be, Grannett and his coauthors[7] wrote, that even if nymphs succeed in establishing themselves on a resistant root, they may not long survive because the root contains toxins or other substances that interfere with the digestion of proteins. But as Grannett points out, there is no direct evidence, such as the isolation and chemical identification of a toxic compound from a resistant rootstock and the demonstration that it kills phylloxeras. The roots are also known to develop a corky layer around a nymph's feeding site, which probably prevents the formation of a gall and consequently the survival of the phylloxera nymph. These possible mechanisms in grape rootstocks are suggested, but not established, by correlations between the fate of the insects and the presence in the plant of chemicals that might be detrimental to the insects.

Survival of the fittest. To achieve evolutionary fitness, any organism—be it plant, animal, or bacterium—must fulfill the three imperatives of existence: it must eat and grow, it must keep from being eaten, and it must reproduce itself. Many aspects of the anatomy, physiology, and behavior of all living things were and continue to be shaped by the struggle to survive until they have produced offspring. The need for plant-feeding insects to eat and the opposing need for plants to protect themselves from being eaten have locked these organisms in continuing cycles in which plants evolve defenses against insects and insects evolve ways of circumventing the plants' defenses.

Riley's rescue of the French wine industry from destruction by the grape phylloxera and the subsequent development of resistance by the phylloxera gives us, from both a historical and an evolutionary perspective, another exceptionally instructive and fascinating view of just one plant/insect evolutionary arms race in action—similar to that of the white cabbage butterfly. But virtually all green plants are fed upon by insects, and they have evolved many different and often amazing defenses. Insects have in turn responded with equally wonderful countermeasures.

Many plants have thorns, hairs, or other physical defenses, but most rely mainly on biochemical defenses; they contain toxic or nasty-tasting chemicals, the latter probably just a warning of the presence of the former. But what are these defensive chemicals? Broadly speaking, as we have seen, plants contain *primary substances* and *secondary substances*. The primary substances, which include nutrients such as proteins, carbohydrates, and vitamins that are required for growth and are essential to all plants and animals, do not, of course, deter insects. The *secondary substances*, a diverse array of tens of thousands of chemicals, only a few of which occur in any one plant, have no known role in the plant's growth or metabolism. Nevertheless, secondary substances are essential to plants. Some benefit both plants and insects, such as the odors—many of them pleasant to the human nose—that attract pollinating insects to nectar-producing blossoms, but most of these secondary substances are "chemical warfare" agents directed at insects. Not surprisingly, people have used several of them as insecticides, notably nicotine from the leaves of tobacco, pyrethrum from the blossoms of certain chrysanthemums, and rotenone from the roots of a South American shrub. Since plant toxins affect the physiological functions of animals, it is not surprising that quite a few of them have medicinal properties: aspirin from willow bark, quinine from the bark of the cinchona tree, digitalis from the garden flower foxglove, and several, including taxol from the yew and vincristine and vinblastine from a periwinkle, that are used in chemotherapy for cancer. Many secondary substances are familiar to us because, although they may be toxic to insects, they give distinctive flavors and odors to vegetables and fruits such as celery, cabbage, and cherries, and to spices such as vanilla, pepper, chilis, and thyme. The distinctive flavor of almonds is due to a low concentration of cyanide.

Herbivorous insects evolve various ways of circumventing the chemical defenses of plants. They may, for example, come up with an enzyme that detoxifies a nasty chemical or find a way of storing it in places in their bodies where it can do no harm. Plants whose defenses have been breached must evolve other defenses or face extinction, and the insects—if they are to survive—must then overcome the new defenses or adapt to another food plant. Thus, the escalating arms race goes on and on, and the number of defensive chemicals synthesized by plants steadily increases.

There are tens if not hundreds of thousands of different plant-insect chemical interactions. It goes without saying that the great, great majority of them have not been subjected to scientific analysis. But a number of research projects, reviewed in Gerald Rosenthal and May Berenbaum's volume on insect-plant interactions,[8] have elucidated in detail some of the many ways in which plants protect themselves against insects and insects get around these plant defenses.

A case in point are the larvae of a seed weevil (*Caryedes brasiliensis*) of the New World tropics that live within and feed on seeds of a shrub (*Dioclea megacarpa*) of the pea family. These seeds contain huge amounts of L-Canavanine, which is known to be highly toxic to most insects because it interferes with the formation of proteins from their building blocks, amino acids. The structure of L-Canavanine is so similar to that of the amino acid L-arginine, a component of proteins, that the protein-forming mechanisms of most seed-eating insects cannot tell the difference and are deceived into producing nonfunctional proteins, usually with lethal consequences, by incorporating in them L-Canavanine rather than the required L-arginine.

But the *Caryedes* seed weevils are not harmed by L-Canavanine because evolution has modified their mechanism for synthesizing proteins from amino acids so that it, the mechanism, can distinguish between L-arginine and L-Canavanine and reject the latter, thereby saving the weevil's life. Furthermore, Gerald Rosenthal and his colleagues[9] found that the weevils have actually turned the tables on the plant by using L-Canavanine as a nutrient, metabolizing it as a source of nitrogen, an essential component of all amino acids and proteins. This is a very good deal for the weevils because 55 percent of all the nitrogen in a *Dioclea* seed is contained in its L-Canavanine.

Many species of plants of the parsley family (Apiaceae) have an unusual and especially interesting chemical defense against plant-eating insects. They are protected by chemicals known as psoralens, which are said to be photoactive, because they are highly toxic when exposed to sunlight but not in the dark. (Some medications, such as tetracycline and St. John's wort, are photoactive, heightening sun sensitivity and causing exceptionally severe sunburn.) Xanthotoxin, a psoralen present in the leaves and/or seeds of many members of the parsley family, absorbs energy from sunlight, especially from its energy-rich ultraviolet rays, and thereby becomes very toxic. It kills susceptible insects by irreparably damaging their DNA, which bears the genetic information that would make possible their normal development.

Some few insects, notably two pests of vegetable crops of the parsley family, the parsleyworm and parsnip webworm, are specialists that have evolved physiological defenses against xanthotoxin and other psoralens. Both of these "worms" are actually caterpillars: the parsley worm the larval stage of the black swallowtail butterfly, and the parsnip webworm the larva of a moth. G. Wayne Ivie and his coresearchers[10] discovered that parsley worms rapidly absorb xanthotoxin into the wall of the gut and there split the xathotoxin molecule into two compounds that are not toxic.

Plants of the parsley family may or may not contain xanthotoxin or other photoactive toxins. Those that do, such as poison hemlock and cow parsnip, grow in open areas with abundant sunlight to activate their psoralens. But nature is ever parsimonious. Those that grow in the shade, as May Berenbaum[11] discovered, do not waste energy formulating psoralens. Among these shade-loving plants are golden

Alexander and sweet ciciley. Most insect generalists, species that feed on many different species of plants, cannot cope with psoralens. The fall armyworm, the larva of a moth, which has a wide-ranging appetite, feeding on corn, beans, potato, cabbage, tobacco, and many other plants of many different families, does have a very limited ability to detoxify psoralens, but not enough to save itself from death. The black swallowtail detoxifies xanthotoxin 50 times faster than does the fall armyworm.

When Riley saved the French grapes from destruction, modern synthetic insecticides that kill insects on contact, such as DDT, chlordane, parathion, and carbaryl, were still far in the future. As Riley might well have foreseen, these "miracle insecticides" are a mixed blessing. Not only do some of them adversely affect nontarget organisms such as beneficial insects and birds, but most, if not all, of them can "create" new pests by killing parasites and predators that control populations of potentially destructive insects that are not killed by these insecticides. Time after time in crop after crop, synthetic insecticides have made pests out of once innocuous insects. This is what happened when DDT and a succession of other insecticides were used to control the codling moth, which in its caterpillar stage is the infamous worm in the apple.

14

AN INSECTICIDE "CREATES" NEW PESTS

Codling Moth

The little moth emerged from her cocoon the previous night and has been sitting on the trunk of an apple tree throughout the daylight hours. Her gray and brown front wings, folded back flat to cover the hind wings, blend in with the bark, a wonderfully deceptive camouflage that hides her from creatures, mainly birds, that would like nothing better than to make a meal of her. After dusk, if the evening is warm, she releases a volatile sex attractant pheromone that, carried by the breeze, will summon a male who will mate with her, providing the sperm that will fertilize the forty or fifty eggs she will lay during her lifetime of two to three weeks. She carefully places those eggs, one at a time, on the upper side of a leaf or twig near a cluster of apples, which are green developing fruits if it is spring or larger ripening fruits in the fall.

Depending upon the ambient temperature, the eggs hatch in as few as six days or, if it is cold, after as many as twenty days. The tiny caterpillars crawl to an apple and burrow into it, sometimes through

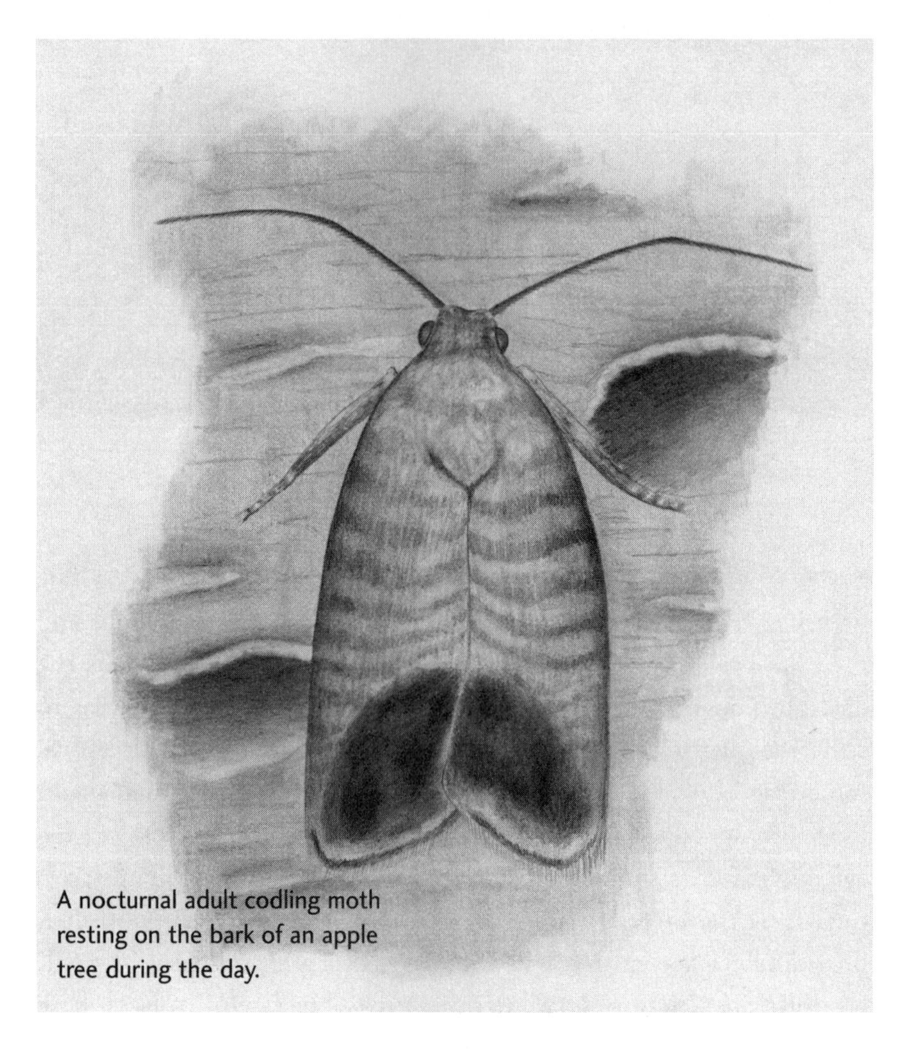

A nocturnal adult codling moth resting on the bark of an apple tree during the day.

the side but more often through the dimple at its blossom end, the end opposite the stem. They tunnel to the core of the fruit where they feed on the seeds and other tissues. Because the seeds contain a large amount of cyanide—you can taste the bitter almond flavor of the cyanide if you crack a seed—most insects are poisoned if they eat them, but these apple-infesting caterpillars can safely eat them because they have enzymes in their digestive system that detoxify cyanide.

In as little as three weeks, the caterpillars are full-grown and tunnel out of the fruit to spin a cocoon, most often under a loose flake of bark on the trunk of the tree, but sometimes in debris on the ground. In all but the most northern of the apple-growing areas, such as Nova Scotia, there is more than one generation per year. In central Illinois there are two and sometimes a partial third, but in very warm areas there may be three or more. In spring the activities of this insect are nicely timed with the seasonal development of apple trees. After a long winter the adults emerge from their cocoons and lay eggs when the trees are in full blossom, and by the time the eggs hatch the trees bear the small, developing fruits required by the caterpillars. In central Illinois, larvae of the first generation grow from egg to adult without interruption, but most second-generation larvae stop developing and go through the winter diapausing in their cocoons. They do not pupate and metamorphose to the adult stage until the following spring.

This insect, the infamous worm in the apple, is the codling moth, *Cydia pomonella*. In *Nature Wars*, a perceptive and eminently readable commentary on the benefits and dangers of the modern practices of pest control, Mark Winston[1] relates that the codling moth was bestowed with its common name in England, where it commonly infests the long, tapering variety of cooking apples known as "codlings."

The codling moth probably evolved in western Asia, but trade and the shipping of fruit has spread it to most of the temperate zone fruit-growing areas of the world. This species infests several fruits other than apple, among them pears, quince, crab apple, and even the fleshy outer husk of the English walnut. Robert L. Metcalf and Robert A. Metcalf[2] noted that the codling moth "is the most persistent, destructive, and difficult to control of all the insect pests of the fruit of the apple. . . ." If left uncontrolled, it may damage and render unmarketable from 20 to 95 percent of the apples in an orchard!

Codling moths have many enemies. The larvae of several species of wasps live as parasites in the bodies of the caterpillars, and one tiny species, a member of the family Trichogrammatidae, lives within the egg.

Bats chase the moths at night, and many insectivorous birds eat them in daylight if they can find them. Even the larvae in their cocoons are not safe. On a winter day I watched a downy woodpecker drill through a flake of bark on an apple tree. After the woodpecker left, I found that the flake had been pierced by a small hole that went dead center into a codling moth cocoon. There is no doubt that the woodpecker had eaten the caterpillar. The cocoon was completely empty. There was no pupal skin, telling me that the insect had not disappeared from the cocoon in the only other way possible, by emerging as an adult the previous summer. Many cocoons on that tree and others nearby had been similarly attacked, but as far as I could find, woodpeckers had not drilled through any bark flakes that did not cover cocoons. How they can tell whether or not a bark flake covers a cocoon remains a mystery. My guess is that the woodpecker taps on bark flakes at random and drills only if it perceives the movements of a caterpillar disturbed by the tapping.

These parasites and predators do not, however, kill enough codling moths to keep the number of damaged fruits below the economic threshold, the point at which the cost of chemical control is equal to or less than the monetary loss to an insect. Because their profit margin is very low, commercial orchardists consider a 0.5 to 1 percent infestation enough to justify the cost of an insecticide application. Consequently, the trees must be sprayed frequently to control codling moths and other insects, and in a few areas so frequently that the cost of control became so prohibitive it put apple growers out of business.

Insecticides are certainly very efficient killers of codling moths and other insects, but they have side effects that may diminish or even wipe out any benefit gained from using them. In other words, an efficient killer of insects is not always a good preventer of the injuries and losses caused by insects.

An important shortcoming of insecticides, which is greatly exacerbated by their overuse, is that many insects become resistant to them, just as house flies, once extremely susceptible to DDT, soon became so resistant to it that they now survive massive doses. Over the years, the overuse of insecticides has decreased, but they are still too often overused. A case in point: some, but not all, corn growers, spurred on by the chemical industry, use an insecticide as a form of "insurance" against corn rootworms, applying it routinely whether or not they are actually faced with a threat from these insects. That is equivalent to taking an antibiotic every day of your life just in case you pick up a germ. Whether or not there is a threat from rootworms could have been determined the previous year by scouting cornfields in which corn would again be planted the following year.

Today 500 or more formidable pest insects, from codling moths to malaria-transmitting mosquitoes, are resistant to one or more insecticides. When resistance develops, other insecticides—usually more expensive and sometimes more hazardous—must be substituted. Codling moths, as noted by Winston,[3] are an instructive example. They have been sprayed with a sequence of several insecticides as they have become resistant to one after another. Before DDT was available, lead arsenate was virtually the only insecticide used to control codling moths. In the mid-1940s, DDT was substituted for lead arsenate but was effective for only a few years. And then, because of resistance, one synthetic organic insecticide after another was substituted, and today apple growers control codling moths with Guthion, a very hazardous chemical related to the infamous nerve gases, some of which are chemical warfare agents. Winston told me that although growers wear gas masks and protective clothing when they spray Guthion, they sometimes become so sick they have to get off their tractors to vomit. Moreover, there are now signs that codling moths are becoming resistant to Guthion. The general public knows little or nothing about resistance to insecticides, but are becoming familiar with a similar problem in the practice of medicine as the news media report case after case of disease-causing bacteria becoming resistant to antibiotics, a situation exacerbated by the notorious overuse of antibiotics.

Many people—as witnessed by the burgeoning of "organic farming"—are concerned about the fact that all synthetic insecticides are more or less poisonous to people and are a threat not only to the grower who applies them but also to the consumer who eats produce that is contaminated with a persistent residue of an insecticide. In 1985 the American public was rudely reminded that insecticides can be a serious threat to their health. In that year, some commercial watermelon growers in California, ignoring the advice of the manufacturer, applied an insecticide called aldicarb to their crops. As a result, thousands of people in the West became ill, some seriously. But this is a relatively minor incident compared to what happened when lead arsenate was used to control codling moths in the late nineteenth and early twentieth centuries. While DDT is stored in our body fat and may have long-term chronic effects such as causing cancer, it is not acutely toxic. In other words, no one has suddenly dropped dead from eating an apple contaminated with DDT. Lead arsenate, on the other hand, is acutely as well as chronically toxic; people can become ill or even die shortly after eating contaminated food.

In October 1925, according to Thomas Dunlap,[4] an analytical chemist in England traced two cases of arsenic poisoning to the consumption of apples imported from the United States. This raised a furor and for a while jeopardized the exportation of American fruit to England. But the British, recognizing that it was impossible to keep apples completely free of arsenic, agreed to continue importing American apples as long as they were contaminated with no more arsenic than what they considered to be a safe dose. They set a "tolerance level" of 1.4 ppm (parts per million) of arsenic. That is, they were willing to tolerate 1.4 parts of arsenic per one million parts of fruit.

Some American orchardists, intent on repairing their reputations and keeping their export market, met the British standards by mechanically washing their apples, thereby removing much of the residue of arsenic. However, there was no maximum tolerance level for apples sold in the United States, though it was imperative that one be set. It would have been insupportable to regulate export shipments without making some provision for safeguarding Americans. Some apples intended for domestic consumption retained a dangerous residue of arsenic, almost 16 ppm, which sickened guinea pigs when they ate peels from the fruit. Objecting to the expense of washing their fruit, western apple growers said that they could not meet the British standards. Under pressure from growers and their congressional allies, the Bureau of Chemistry of the USDA, later to become the Food and Drug Administration, set a too high tolerance of 3.6 ppm on apples for domestic consumption. Tolerating contamination with any amount of arsenic did not take into account the known fact that even tiny amounts of arsenic accumulate in the body and ultimately have life-threatening chronic effects.

In 1935 Walter G. Campbell, commissioner of the Food and Drug Administration, testified before Congress on the dangers of using arsenic as an insecticide. According to Dunlap,[5] Campbell told of several cases of poisoning and the deaths of two children, offering supporting evidence from physicians and autopsy results. But he stressed that the "chief danger was not acute poisoning, but the cumulative deterioration of liver, kidneys, and other organs by the continued consumption of small amounts of lead and arsenic."

By 1932 the tolerance for arsenic on fruit for domestic consumption had been lowered to the British level of 1.4, but in 1940, at the insistence of the growers, it was raised back to 3.6 ppm. Dunlap writes that no one claimed that the higher levels would improve public health, "only that the public apparently could stand them and the farmer profit from them by spraying with less caution and cutting the expense of cleaning his crop."[6]

Insecticides have another insidious repercussion. They often bring about unforeseen and unwelcome ecological effects by killing insects other than the target species. A widely known and well-understood example is the veritable reorganization of the apple orchard ecosystem —its web of life—caused by the use of DDT and the other synthetic insecticides later used to control codling moths. As improbable as it seems, these insecticides can create new insect problems (secondary pests) by making it possible for insects that were originally too scarce to cause significant losses to become so abundant that they pose serious if not potentially devastating threats.

In 1946 DDT was first used in apple orchards. Until codling moths became resistant to this insecticide, it did a good job of controlling them if it was sprayed on the trees once every two weeks. Only six years later, in 1952, E. H. Glass and P. J. Chapman of the New York State Agricultural Experiment Station said of the red-banded leaf roller (a native caterpillar that feeds on both apple leaves and fruits) that prior to 1946 it caused occasional moderate losses in apple orchards in the northeastern United States, but was "generally rated . . . as a minor pest."[7] However, "the economic status of the red-banded leaf roller changed rather abruptly about 6 years ago [in 1946 when DDT was first used]. It then became a major pest and is now a serious enemy of the apple in the principal fruit-growing districts of the eastern United States." They suggested that the upsurge of the leaf-roller population was caused by the use of DDT. Because of the new threat from this insect, two more insecticides were added to the spray schedule, DDD, an analogue of DDT, and parathion, another compound in the same chemical group as the nerve gases.

A year later, in 1953, C. R. Cutright of the Ohio Agricultural Experiment Station reported that the European red mite, not an insect but an eight-legged member of the group that includes the ticks and spiders, became a serious pest of apples in 1946, coincident with the first use of DDT: "In 1944 and 1945 when lead arsenate was the main insecticide only one orchard [of fifteen] . . . was infested to a serious extent by the European red mite."[8] But in 1946 eight of ten orchards treated with DDT suffered commercially significant damage from this mite, which sucks sap from the leaves, injuring them so severely that the tree bears undersized apples of poor quality and color. As the Metcalfs put it, this mite, the two-spotted spider mite, and three others are all "insecticide induced secondary pests."[9] The upsurge of mite populations necessitated adding yet another chemical to the spray schedule, an acaricide, a killer of mites (from the Greek *acar*, "a mite," and the Latin *cide*, "kill"). As many as seven additional pesticide applications were used to control just these mites. Some orchards were sprayed 20 times in one season, including sprays for mites, red-banded leaf rollers, and the usual pests such as codling moths.

But how do DDT, and now other synthetic insecticides, "create" new pests of apples or other crops? DDT, for example, provided excellent control of the codling moth until it became resistant to this insecticide, but it was not an effective control for either the red-banded leaf roller or the apple-infesting mites because it is much less toxic to them. In 1952 Glass and Chapman recognized "that conditions were made highly favorable for leaf-roller activity by the use of DDT through the virtual elimination of its natural enemies. . . ."[10] The toxicity of DDT to the parasites and predators that are the natural controllers of leaf-roller and mite populations greatly aggravated a minor problem, upsetting the balance of the orchard ecosystem to the point that an upsurge of these once almost inconsequential bugs blazed out of control.

Although lead arsenate is toxic to them and all other animals, predaceous and parasitic insects were not affected by it. The arsenicals are stomach poisons. That is, they have no effect unless they are ingested. Red-banded leaf-roller caterpillars and other chewing insects swallow lead arsenate when they feed on the leaves or the outer skin of fruits. But parasites and predators, which do not eat the leaves or fruits, do not ingest lead arsenate and are not harmed by walking on it. However, they are killed by DDT and most of the other modern insecticides, which are *contact poisons* that are absorbed through the skin of their feet or other body parts just as they are absorbed through the skin of codling moths.

In 1974 Ronald Meyer[11] described an ingenious way of minimizing the use of pesticides in apple orchards by managing orchards so as to maximize populations of the predaceous fallacis mite, which eats plant-feeding mites such as the European red mite and the two-spotted spider mite. This is a spectacular and instructive example of the modern concept of *pest management*, which seeks to minimize—but not necessarily eliminate—the use of pesticides. The idea is to combine all available methods of control—such as using crop rotation, choosing a propitious planting date, encouraging parasites and predators, and making judicious use of pesticides—so that economic damage is avoided and adverse side effects such as contamination of the environment are at least minimized.

European red mites survive the winter as diapausing eggs in crevices in rough bark on the undersides of twigs high in the apple tree. The eggs hatch just before the trees blossom, and the minute nymphs crawl to nearby leaves to suck sap, grow, and reproduce. Because they have a very short life cycle—developing from egg to reproducing adults in as little as four days—they can, if left uncontrolled, quickly become numerous enough to do serious damage. Early in the season one application of a pesticide is usually required to suppress European red mites, but only this one spray is needed if the orchard is managed so that the predaceous fallacis mites are abundant.

Fallacis and two-spotted spider mites overwinter only as mature females hiding at the base of the apple tree's trunk: under loose bark; among apple shoots, grasses, or other small plants; or in debris on the ground. At about the same time that the eggs of the European red mite hatch up in the tree, the overwintered female fallacis and two-spotted spider mites become active and begin to lay eggs at the base of the tree. The spider mites survive by sucking sap from almost any of the ground-cover plants, and the fallacis mites survive by preying on the spider mites. Both species multiply and the population of the predaceous fallacis mites spreads upward throughout the tree and controls the European red mites and all other plant-feeding mites for the rest of the season.

This system works only if a plentiful population of plants ("weeds"), the nursery for the fallacis mites and their prey, is left to grow at the base of the apple trees. Bill Luckmann told me that a meticulously groomed orchard with no plants flourishing at the base of the trees is a "sorry sight." Fortunately, the fallacis mite has become resistant to the acaricide used to control European red mites early in the season, and to the several insecticides routinely used to control codling moths and other apple pests. Resistant insects or mites are usually a problem, but in this case the resistant fallacis mites are the solution to a problem.

The ecological backlash caused by the unwise use of insecticides in apple orchards is only one of many such self-defeating attempts to control pest insects with DDT, parathion, carbamates, or other broad-spectrum insecticides that indiscriminately kill many different kinds of insects, parasites, and predators as well as the target plant-feeders. A few examples make the point. The biological control of the horrendously destructive cottony cushion scale in California citrus orchards by its insectivorous enemy, the vedalia beetle (more about this in chapter 19) totally failed when an insecticide that kills vedalias but not cottony cushion scales was applied to control another pest of citrus. The biological control was restored when the use of this insecticide was discontinued.

When DDT was sprayed on elm trees to kill the bark beetles that transmit Dutch elm disease, scale insects and aphids, previously uncommon and of little significance, became distressingly numerous and destructive to the elms, because DDT, not toxic to them, killed their parasites and predators. In other places, the disease was controlled by cutting and burning the dead and diseased branches and trees in which the bark beetles breed.

By 1970 cotton production in the Lower Rio Grande Valley in northeastern Mexico was abandoned because insecticides to control the boll weevil, the primary pest of cotton, created destructive secondary pests not susceptible to these insecticides by destroying their natural enemies. Other insecticides applied to control these secondary pests created yet more secondary pests that had to be controlled with yet more new insecticides that created even more secondary pests. This "pesticide treadmill" continued until the growers were treating their fields with insecticides an incredible 18 times per season and still suffered crippling losses in yield. By 1970 the growing of cotton was abandoned in this area because of crop losses and the prohibitive cost of insecticide treatments.

The European corn borer, true to its name, came to us from Europe and bores in the stalks and ears of corn plants. The weak point in its life cycle, an opportunity for controlling it, is that it spends the winter as a caterpillar in a burrow in a corn plant. It emerges from the dead plant as an adult moth in spring through an escape tunnel that it had carved when it was still a larva. Ideally, corn borers could be controlled by plowing down cornstalks before spring, thereby blocking the escape tunnels with soil and thwarting the emergence of the adult moths. However, this never worked in practice because of a succession of incompatible agricultural practices. Farmers have, therefore, relied on insecticides. But in recent years some farmers have used genetically modified (GM) corn plants to control corn borers. A bacterial gene for the synthesis of a naturally occurring insecticide, B_t, has been inserted into these plants. Although B_t is harmless to people, this practice—as has been the use of all genetically modified organisms (GMOs)—has been controversial.

15 FROM LOW- TO HIGH-TECH CONTROLS

European Corn Borer

"A research worker of the Massachusetts Agricultural Experiment Station in 1917 discovered several pinkish-brown worms on sweet corn in market gardens near Boston. Specialists examined the larvae and found them to be a species that was a pest of corn in Europe. A bit of sleuthing disclosed that they had sneaked into this country a few years earlier in broomcorn imported probably from Italy or Hungary for use in broom factories in Medford, Massachusetts." So wrote William Bradley,[1] director of a crop pest research laboratory, of the moth known on this side of the Atlantic as the European corn borer.

The European corn borer is one of the many significantly destructive pest insects that, like the gypsy moth, have been unintentionally brought into North America from abroad and have become established here. The introduction of the plant-boring European corn borer was a calamity for North American agriculture. It is very destructive to both field and sweet corn and eventually infested all of the major corn-growing areas of the United States and southern Canada, spreading rapidly from its foothold in New England. By 1938 it occupied most of the northeast, by 1948 most of the Midwest as well, and today its range extends from Canada to the Gulf States and west to the Rocky Mountains.

Despite their common name, European corn borers (*Ostrinia nubilalis*) feed on many different kinds of plants. By 1927 D. J. Caffrey and L. H. Worthley[2] had already listed over 200 plant species on which these borers feed in North America, including many wild plants and some cultivated ones such as potatoes, bell peppers, celery, and gladioli. But it is most often found infesting corn.

Corn borers spend the winter as full-grown, diapausing caterpillars hidden in a tunnel or cavity they excavate in the stalk or in an ear. Before retiring for the winter, they excavate an escape tunnel that is usually loosely plugged with fecal pellets. Diapause is terminated by the long days of spring, as D. G. R. McLeod and Stanley Beck[3] discovered. But unlike many other insects, a period of chilling is not a prerequisite to the termination of diapause. After the reactivated caterpillar spins a flimsy cocoon in its cavity, it molts to the pupal stage, and in about two weeks the adult moth sheds its pupal skin. The "toothless" moth, which has no mandibles and could not chew its way out of a cornstalk, makes its way to the outside world through the escape tunnel.

Large numbers of adults of both sexes gather in densely vegetated areas such as fencerows or ditch banks, refuges where they can find a humid and otherwise favorable microclimate. Between 10:00 PM and 1:00 AM, females attract mates by releasing a sex pheromone into the air. Between sundown and midnight on temperate evenings, mated females lay small masses of white, flattened eggs, which overlap like fish scales, on the underside of corn leaves. Then they return to their refuge and refresh themselves with dew drops that collect on plants. They are likely to continue laying for as long as 10 days, and a female can produce an average total of 500 or sometimes many more eggs.

In about a week the eggs hatch. Young larvae feed on the leaves of

Female European corn borer moth and eggs that she just laid on a corn leaf.

the corn plant, but when they are about half grown they become borers, burrowing and eating their way into the thick part of a leaf stem, the stalk, the shank of an ear, or into the ear itself. This feeding alone can significantly reduce yield, but can also lead to more serious losses by causing tassel stems to break, ears to drop off, or whole plants to fall over because their weakened stalks break. In much of the Midwest, including central Illinois, corn borers have two generations per year. The spring generation completes its metamorphosis without the intervention of diapause, but larvae of the summer generation go into diapause in response to the short days and relatively low temperatures of late summer and early autumn.

American agricultural entomologists soon realized that corn borers might be controlled by clean plowing, destroying the overwintering larvae by plowing down and burying the cornstalks and other debris that houses them—a good idea in theory, but not in practice. When the corn borers first arrived, midwestern farmers were growing a lot of oats, which were usually sown in fields with corn stubble that had not been plowed under after the harvest, only broken up with a harrow. Many corn borers survived in these fields and were a source of infestation. This particular impediment almost disappeared when soybeans, which in those days were always planted in plowed ground, almost completely replaced oats and most crops other than corn. Even so, clean plowing would not have solved the corn borer problem even if all farmers had meticulously plowed down all corn debris before the mother borers emerged in spring. The reason for this, discovered by H. C. Chiang[4] of the University of Minnesota, was that many corn borers survived the winter in the cobs of whole ears of corn stored during winter in "cribs" from which the moths could easily escape in spring. The current use of a new harvester, the picker-sheller, which removes only the kernels from the cob and leaves all corn debris, including cobs, in the field to be plowed down, might have averted this problem, except that many farmers have adopted a system of minimum tillage that impedes erosion by leaving cornstalks and other crop debris on the surface of the soil. Obviously, some other way of controlling corn borers needs to be used.

If a field of corn is seriously threatened by corn borers, the proper application of an insecticide at the right time may avert a significant economic loss. But what constitutes a threat that justifies the use of an insecticide, and how do you determine the right time to apply it? The answers to these questions depend upon the growth stage of the plant. Young corn plants are very resistant to corn borer larvae, because the plants contain a toxin whose technical chemical name is a monumental jawbreaker: 2, 4-dihydroxy-7-methoxy-1, 4-benzoxazine-3-one. To the relief of everyone who has to talk about this chemical, its name is abbreviated to the acronym DIMBOA, which is pronounced as a word. Young corn plants with a large content of DIMBOA are, according to Dean Barry and his coauthors,[5] very resistant to newly hatched corn borer caterpillars, which refuse to feed on the leaves and die, sparing the plant the serious damage they would do after they grow larger.

As corn plants mature physiologically, the concentration of DIMBOA decreases until they are no longer resistant to corn borers. But how do you determine the physiological age of a plant, which is, of course, a measure of its DIMBOA content? It would be a waste of time and money and an unjustifiable contribution to environmental pollution to apply an artificial insecticide when the plants are still protected by their built-in natural insecticides. (But it is a sad fact that few farmers use the system described below.) The chronological age of the plants is not a reliable indicator of physiological maturity, because the rate at which corn plants mature varies with environmental conditions and their horticultural variety. The same can be said for the height of the plant, which was once used as an index of maturity. Some varieties are tall and others are short, and height can also vary with environmental conditions.

Two entomologists at the Illinois Natural History Survey in Champaign, William Luckmann and George Decker,[6] devised an index, "the tassel ratio," which is closely correlated with the state of a corn plant's physiological maturity and is, therefore, a good measure of whether or not its resistance to corn borers has diminished or been lost. They did this more than a decade before Stanley Beck[7] and his group discovered that it is DIMBOA that protects young corn plants against corn borers. Although Luckmann and Decker did not know why corn plants are resistant, their tassel ratio is, nevertheless, a measure of a corn plant's DIMBOA content.

The tassel of a flowering corn plant is a male structure that produces copious quantities of pollen that will be blown by the wind and may, with luck, land on and adhere to the silks of an ear, thereby fertilizing this female part of the plant. On a sexually mature plant, the plumelike tassel rises above the rest of the plant on a thin stem. Before the tassel becomes externally visible, it is hidden within the stalk, growing ever upward until it emerges from the whorl of leaves at the top of the growing plant. The developing tassel can be revealed by slitting the stalk lengthwise down its middle. For about the first 30 days after germination it is not visible, but on the thirty-eighth day Luckmann and Decker split a stalk, revealing a tiny tassel just beginning to develop and less than two inches long.

To find the tassel ratio you first extend the longest leaf straight up and measure the distance from its tip to the base of the stalk at ground level; this is the extended plant height. Then you cut the plant down at ground level, split the stalk, and measure the tassel height. The tassel height divided by the extended plant height and multiplied by 100 is the tassel ratio.

Luckmann and Decker[8] established that chronological age is indeed not a reliable indicator of maturity. They found no correlation between the chronological age of plants and their level of maturity as measured by the average tassel ratio of a randomly selected sample of several plants from the same field. In 1949, for example, a certain variety had an average tassel ratio of 48 when the plants were 46 days old, but in 1950 plants of the same variety had a tassel ratio of only 24 when they were of the same chronological age. Nor was there a correlation between extended plant height and tassel ratio. In three succeeding years, plants with an extended plant height of 48 inches had tassel ratios of 12, 14, and 23.

What does the tassel ratio tell us about when or whether or not to apply an insecticide? During the first 30 days after they germinate, when the tassel has not yet begun to develop, corn plants are immune to the attack of European corn borers. It is not until tassel ratio 20, six or more weeks after germination, that a few young borers, a minor threat, manage to establish themselves on a plant. Bill Luckmann told me that before tassel ratio 20, newly hatched corn borers take a few test bites from a leaf and then inevitably move away and descend to the ground, where they are eaten by birds, ants, or other predators.

But again, when should an insecticide be applied and how big a threat from corn borers requires intervention with an insecticide? According to William Luckmann and Howard Petty,[9] an insecticide need not be applied until tassel ratio 30. The rule, based on many field trials, is that in a field with an average tassel ratio between 30 and 50, the application of an insecticide is useful, but *only* if 75 percent or more of the plants show fresh leaf feeding. But by tassel ratio 60, when the tassels first begin to emerge from the stalk, it would be useless to apply an insecticide, because by then the borers will have tunneled into the plant where they are shielded from insecticides.

The next chapter in the saga of European corn borer control was and still is the use of genetic engineering to add new weapons to the corn plant's natural defenses, weapons that would make the plant resistant to insect attack as does DIMBOA, but in all its growth stages. The plants have been made more resistant by using the modern methods of biotechnology to insert into their DNA two toxin-producing genes from the soil-dwelling bacterium *Bacillus thuringiensis*. The gene inserted in corn produces a protein (B_t) toxic only to caterpillars of moths and butterflies. Other strains kill other insects; the strain *israelensis*, for example, kills mosquitoes. When a corn borer feeds on a B_t corn plant, it ingests the toxin, which causes cells in its gut to rupture, making a hole in the gut's lining. As a result, the borer may be paralyzed, stop feeding, and starve to death, or if it is not paralyzed, it will be killed by the runaway multiplication of the bacteria that are normally confined to its intestines.

Corn is just one of several genetically modified organisms (GMOs). Soybeans have been inserted with a gene that makes them resistant to Roundup, a powerful herbicide that can then be used to kill weeds without doing serious harm to the genetically modified soybeans. Cotton containing the B_t gene is resistant to the cotton bollworm. Hemophiliacs are helped by blood-clotting factors produced in the milk of pigs or goats into which the appropriate human gene has been inserted. These clotting factors were once extracted from human blood, but that is less efficient and is too risky because of the possibility of contamination with HIV and the hepatitis C viruses. Large quantities of the insulin required by diabetics is produced by cultures of bacteria into which the human gene for insulin production has been inserted.

GMOs can surely be a blessing, but it has become apparent that at least some of them are a mixed blessing. Many contend, and I think rightly so, that they have not been adequately tested for consumer and environmental safety. For this reason, many Europeans will not eat foods made with GMOs, a serious problem for American farmers because much of their corn is exported—40 percent of the Illinois crop alone. Some proponents of GMOs have argued that there is really no need for concern because plant breeders have for 200 years or more been transferring genes from one kind of plant to another by selective breeding and hybridization. I do not find this argument reassuring, because no conventional breeder has ever managed to transfer genes between two such very distantly related organisms as a bacterium and a corn plant or a human and a pig. The question of the effects of GMOs on human and animal health has not been completely answered and is still being debated.

There is, however, good evidence that at least one type of genetically altered corn has had adverse environmental effects by killing nontarget insects—the monarch and the black swallowtail caterpillars—in the field. This story began in 1999 when John Losey and his coworkers[10] in Cornell University's Department of Entomology did a laboratory assay which showed that over 50 percent of a group of monarch larvae died after feeding for four days on milkweed leaves that had been dusted with pollen from corn plants with the B_t gene. This information received a great deal of attention from the media, arousing the public and alerting them to the environmental damage that might be caused by GMOs. The proponents of GMOs immediately began "damage control" by voicing various criticisms of the Cornell experiments. A fair criticism, the one that made the most sense, is that what happens in a laboratory is not necessarily what happens in the field.

But in 2001, Arthur Zangerl, May Berenbaum, and others[11] of the University of Illinois Department of Entomology did the requisite field experiment and backed it up with a laboratory experiment. Their laboratory results leave no doubt that the pollen of at least one type of B_t corn (Max 454) kills black swallowtail caterpillars. They fed newly hatched caterpillars wild parsnip leaves that had been "seeded" with pollen grains from Max 454 corn plants at four densities, from a high of about 64,500 to a low of 64 grains per square inch. At the greatest density, not likely to occur in the field, 85 percent of the caterpillars had died by the fourth day. At the lower density of 645 grains per square inch, much less than the greatest density seen in the field, over 1,100 grains per square inch, almost 30 percent of the caterpillars were dead on the fourth day.

In the field, Zangerl and his colleagues placed groups of potted milkweed and wild parsnip plants at distances of from about 3 to 23 feet from the edge of a field of Max 454 corn. Careful sampling showed that, as expected, the average number of pollen grains that landed on leaves decreased with the plant's distance from the edge of the cornfield. On milkweed leaves there were as many as 710 per square inch at a distance of about 3 feet and only about 64 at 23 feet. They found fewer pollen grains on wild parsnip leaves: 161 per square inch at 3 feet, and only about 64 at 13 and 23 feet.

Zangerl and his researchers[12] placed newly hatched monarch and black swallowtail caterpillars on potted milkweed and wild parsnip plants in the field four days after the Max 454 plants began to shed pollen. The number of surviving larvae was counted each succeeding day for almost a week. Many larvae were found to be missing and presumably dead, but because insectivorous insects, notably the multicolored Asian ladybird beetle, were numerous on the plants, it was not possible to determine how much of the observed mortality was caused by predators or by B_t pollen.

Too few monarch caterpillars survived for further analysis, but it is clear that B_t pollen had affected the growth rate of the black swallowtail caterpillars that did survive. Their weight gain decreased with their proximity to the corn. At the end of the experiment, caterpillars 23 feet from the B_t corn, where there was relatively little pollen on the leaves, weighed three times as much as caterpillars only 1.5 feet from the B_t corn, where the amount of pollen on the leaves was greatest. Recent fieldwork by several others, reported in a compilation of papers, *B_t Corn Pollen and Monarch Butterflies*, edited by May Berenbaum,[13] showed that, like black swallowtail caterpillars, monarch caterpillars feeding on pollen-bearing milkweed leaves from within a field of Max 454 corn gained much less weight and were much less likely to survive than caterpillars fed milkweed leaves from outside the field.

Some argued that there was no threat to monarch caterpillars from B_t pollen because milkweed does not grow close enough to cornfields. Contrary to their claim, milkweed grows *in* cornfields and also very close to them. May Berenbaum told me that wild parsnip may also grow close to cornfields. The time of year when monarch caterpillars are still feeding on milkweed leaves does not, fortunately, overlap everywhere with the time at which corn is shedding its pollen, but the overlap is greater in the northern part of the monarchs' summer breeding range than in the southern part, because the plants shed pollen earlier in the warmer south. Rainfall protects monarchs from B_t pollen. One of the articles in the compilation edited by Berenbaum reported that a single rain can wash from 54 to 86 percent of the corn pollen from milkweed leaves. The monarch population is especially vulnerable because, as Leonard Wassenaar and Keith Hobson[14] found, about half of the monarchs in North America grow up in just a relatively small part of their total breeding range, an area that includes most of the corn belt. And in much of the corn belt, as in my home county of Champaign, Illinois, corn and soybeans cover almost all of the available land, so milkweed plants have few places to grow other than in or near cornfields.

Max 454 is being withdrawn from the market and is being replaced with B_t corn varieties whose pollen contains so little B_t toxin that it does not kill monarch caterpillars at pollen densities that are likely to occur in the field. Although the toxicity of Max 454 pollen is no longer an issue, there is a lesson to be learned from its unexpected environmental effects. It did come on the market, even if only for a short time, and its pollen did kill nontarget caterpillars. Let us hope that this does not happen again.

The misuse of synthetic insecticides such as the now-banned DDT and dieldrin has caused much more serious—actually catastrophic—damage to the environment. As so interestingly and cogently told in Rachel Carson's *Silent Spring* (1962),[15] the Japanese beetle, as we will see, is the central character of the story of an environmental catastrophe caused by the blatant misuse of an insecticide that is dangerously toxic to birds and mammals. It was indiscriminately showered from the air and from the ground on the fields, pastures, woodlands, and homesteads of a large part of Iroquois County, Illinois. This insecticide was used in a misguided and unsuccessful attempt to eradicate the Japanese beetle from that area to prevent it from invading the rest of Illinois. Carson's book, a bestseller, sparked the movement to ban DDT, dieldrin, and related chlorinated hydrocarbon insecticides in the United States as well as much of the rest of the world.

16

THE DEMISE OF DDT

Japanese Beetle

The Japanese beetle was first seen in the New World in a nursery near Camden, New Jersey, in 1916—apparently brought in with the soil on the balled roots of nursery stock from Japan. Fifteen years later, E. Dwight Sanderson wrote that this invader (*Popillia japonica*) "has been confined very closely to this area but now occurs in isolated infestations in neighboring states and will probably establish itself gradually over a much larger territory."[1] He was right. By the summer of 1927 it had arrived in my hometown, Bridgeport, Connecticut, and by the end of the century had followed me to my adopted hometown, Champaign, Illinois.

In 1938, when I was ten years old, gardeners in my Bridgeport neighborhood were paying kids to get rid of the hordes of Japanese beetles that were devouring their plants. Ten cents for every pickle jar full of beetles drowned in water mixed with a little kerosene was the going rate. (Today that doesn't sound like much money, but in those days a candy bar, a Baby Ruth or a Milky Way, cost only a nickel.) Furthermore, it was easy to fill jars because the beetles were so easy to collect. Obviously liking company, a closely clustered group of one or two dozen—or even more—often occupied a single leaf and was easily brushed into a jar with the hand.

Adult Japanese beetles are of medium size and metallic green and bronzy red. They feed on the foliage, flowers, and fruits of over 350 different kinds of plants, but in my neighborhood they seemed to be particularly fond of roses, hollyhocks, and grapes. These beetles, both adults and larvae (grubs) can be distressingly destructive. The larvae live in the soil feeding on the roots of grasses and other plants, often doing serious damage to lawns and golf greens. The adults, which emerge from the soil in June, "skeletonize" leaves, eating most of the soft tissues but leaving intact all but some of the smallest of the tougher veins. As I write, I have before me a large sycamore leaf picked from a tree on the University of Illinois campus in Urbana. A lacy apparition of its former self, it is so sheer that you can read print through it. Many of the leaves on that tree had been similarly damaged. The adults also spoil fruits by gouging them with shallow holes, and they feed avidly on the silks of corn, clipping them from the ear, thus blocking pollination and thereby preventing the development of kernels. Females lay their eggs two to six inches deep in soil covered with vegetation. In a few days the eggs hatch; the larvae feed on roots until October and then remain in diapause in the soil until June, when they pupate and metamorphose to the adult stage.

By 1953 the infestation of Japanese beetles had spread to the western border of Indiana and a small area of adjoining Iroquois County, Illinois. In 1954 the Plant Pest Control Division of the Agricultural Research Service, the US Department of Agriculture, and the Illinois Department of Agriculture began a joint campaign of spraying from the air the insecticide dieldrin, a chlorinated hydrocarbon chemical compound, as is DDT, to prevent the Japanese beetle from invading Illinois. From 1954 to 1958, according to the economic entomologists William Luckmann and George Decker[2] of the Illinois Natural

History Survey, 17,844 acres of Iroquois County in the vicinity of the town of Sheldon were showered with dieldrin. The airplanes treated virtually everything with this dangerously toxic insecticide: crop fields, pastures, woodlands, farmsteads, and the whole town of Sheldon. As we will see, the campaign failed to eliminate the Japanese beetle and was an unmitigated economic and ecological disaster.

In those early days, flush with the excitement engendered by DDT and the other new synthetic, organic insecticides, economic entomologists were eager to use these new and assumed-to-be-safe weapons against pest insects. An anonymous publication of the US Department of Agriculture, *How to Spray the Aircraft Way* (Farmer's Bulletin 2062), enthused that "spray day can be payday for farmers everywhere." Dazzled by the exuberant reports of the new insecticides as panaceas, cure-alls, for insect problems and as the "atomic bombs of insecticides," many but not all economic entomologists shifted their research interests from the basics of their field—understanding insects in order to better control them—to the new "miracle insecticides."

This shift is reflected by the contents of scientific articles published in the *Journal of Economic Entomology*. Of the 78 articles in two issues of the 1940 volume, only 39 (50 percent) concerned insecticides, most of them naturally occurring insecticidal compounds from plants: pyrethrum, nicotine, and derris, which, except for pyrethrum, are quite toxic to birds and mammals, but are readily biodegradable, that is, they rapidly break down outdoors into simpler, nontoxic substances. Twenty years later, 106 (65 percent) of the 169 articles in the corresponding issues of the 1960 volume of the same journal dealt with the new synthetics, most of which are toxic to mammals and birds, and in the case of the chlorinated hydrocarbon insecticides biodegrade very slowly, persisting in the environment for months, years, or even decades. Even today, for example, almost 30 years after DDT was banned in the United States, small residues can still be found in the Great Lakes and the soil of some orchards.

The progress made by applied entomology since the synthetic insecticide craze is made evident by the content of two 2003 issues of the same journals. Of 107 articles, 20—only 18 percent—were about synthetic insecticides, 12 (11 percent) about resistance to synthetics, and 37 (34 percent) about the pest insects themselves. Thirty-eight, a healthy 35 percent, were about alternatives to our dependence on synthetic insecticides. Seventeen of these considered the new and rational concept of "pest management," minimizing the use of insecticides by integrating them with noninsecticidal methods of control, such as those we will soon come to.

The control campaign in Iroquois County, at best the result of wishful thinking, was an ecological nightmare. No wonder, dieldrin is extremely toxic to birds and mammals, 50 times more so than its cousin DDT. If ingested, only 0.0003 ounce will kill half of a group of 8-ounce rats. (Endrin, another DDT relative used as an insecticide, is so much more toxic it was once registered—licensed—as a spray for killing mice in orchards.) In 1954 Thomas Scott,[3] a wildlife biologist, and two of his colleagues at the Illinois Natural History Survey studied the effect of dieldrin on birds and mammals in Iroquois County and found that many different kinds had been killed. "Robins, brown thrashers, starlings, meadowlarks, common grackles, and ring-necked pheasants . . . ground squirrels, muskrats, and cottontails were virtually eliminated from the treated area." Some farm cows and sheep in pastures contaminated with dieldrin spray became sick and died.

Rachel Carson deplored the fact that:

> The full extent of the toll that has been taken by this largely ineffective program may never be known, for the results measured by the Illinois biologists are a minimum figure. If the research program had been adequately financed to permit full coverage, the destruction revealed would have been even more appalling. But in the eight years of the program, only about $6,000 was provided for biological field studies. Meanwhile the federal government had spent about $375,000 for control work and additional

> thousands had been provided by the state. The amount spent for research was therefore a small fraction of 1 per cent of the outlay for the chemical program.[4]

The ecological catastrophe in Iroquois County and similar insecticide-caused debacles elsewhere inspired Carson, a trained marine biologist, to write *Silent Spring*, which appeared in 1962 and soon became a bestseller. She began the book with what she called "A Fable for Tomorrow," although its unhappy forewarning had already come to pass in some places. The fable is about the effects of insecticides on a town she said did not exist, but it prophetically mirrors what happened in real places such as Sheldon.

> There was once a town in the heart of America where all life seemed to live in harmony with its surroundings. The town lay in the midst of a checkerboard of prosperous farms, with fields of grain and hillsides of orchards where, in spring, white clouds of bloom drifted above the green fields. In autumn, oak and maple and birch set up a blaze of color that flamed and flickered across a backdrop of pines. . . . The countryside was . . . famous for the abundance and variety of its bird life. . . .
>
> Then a strange blight crept over the area and everything began to change . . . mysterious maladies swept the flocks of chickens; the cattle and sheep sickened and died. . . . There was a strange stillness. The birds, for example—where had they gone? Many people spoke of them, puzzled and disturbed. The feeding stations in the backyards were deserted. The few birds seen anywhere were moribund; they trembled violently and could not fly. It was a spring without voices. On the mornings that had once throbbed with the dawn chorus of robins, catbirds, doves, jays, wrens, and scores of other bird voices there was now no sound; only silence lay over the fields and woods and marsh.[5]

Silent Spring alerted the public to what was happening and started a heated debate that sparked a long-overdue examination of the then almost completely unknown effects of synthetic chemical pesticides on the environment and nontarget organisms including humans. On the one hand, some scientists, particularly economic entomologists, agricultural interests including the US Department of Agriculture, and, of course, the pesticide industry defended DDT, dieldrin, and the other new chlorinated hydrocarbon insecticides. On the other hand, many scientists, especially ecologists, some members of the medical community, the alarmed public, and people who would today be called environmentalists pointed to the dangers of these supposedly "safe" insecticides and denounced their misuse.

The failed and environmentally catastrophic attempt to eliminate Japanese beetles from Iroquois County was by no means the only ill-conceived chemical insect-control scheme that went sadly awry. In cities and towns and on college campuses from New England to the Midwest, there were attempts to halt the spread of the deadly Dutch elm disease by spraying DDT to kill the elm bark beetles that transmit this fungal disease from sick to healthy trees. Still, they were miserable failures. The disease continued to spread unabated, and robins were killed by the hundreds when they ate earthworms that had ingested DDT with the organic debris on which they fed and had stored large quantities of it in their body fat. This was the first indication that a pesticide could be passed from prey to predator and concentrated as it moved up a food chain. At about the same time, a large population of nesting western grebes, fish-eating birds, were almost totally eliminated from Clear Lake, California, when DDD, another chlorinated hydrocarbon insecticide, was put in the lake to control a troublesome gnat. The dosage applied was too low to harm the grebes directly, but it did do serious harm to them—killing them and preventing them from reproducing—after it was concentrated to a toxic level as it passed up the food chain from single-celled algae, to tiny crustaceans and insects, to minnows, and finally to larger minnow-eating fish that were eaten by the grebes.

Many birds were affected by a chronic rather than the acute poisoning that caused Iroquois County birds and mammals, and also robins elsewhere, to drop dead shortly after being poisoned. In relatively large doses, both dieldrin and DDT kill by interfering with the transmission of nerve impulses, causing severe tremors, paralysis, and imminent death. In the 1950s, when the University of Illinois campus was being sprayed with DDT in yet another futile attempt to control the Dutch elm disease, I sometimes saw dying robins on the quad—sitting and shaking or, when closer to death, lying on their backs with their legs trembling uncontrollably.

However, DDT, or rather its breakdown product DDE, usually has a less obvious and more insidious chronic effect on birds—almost always on those at the top of a food chain. Fish-eating species (grebes, pelicans, ospreys, and bald eagles) and also bird- and mammal-eating ones (kestrels, peregrine falcons, and other raptors) were seriously affected and in some instances all but extirpated. In these chronic cases, the effect was not on the nervous system but rather on the hormonal system responsible for the deposition of calcium in the eggshell. These birds laid eggs with shells so thin that they were broken by the weight of an incubating parent.

Desperate to "defend their turf," proponents of DDT made the unsubstantiated claim that eggshells had always been that thin. But this absurd notion was soon refuted by comparing the thickness of the shells of recently laid eggs with those of museum specimens that had been collected long before the advent of DDT.

There was an indisputable correlation between eggshell thinning and the presence of DDE in the body of the mother bird. But the proponents of DDT pointed out—quite correctly—that a correlation does not prove cause and effect, that eggshell thinning might be caused by some unknown factor other than DDE. Wildlife biologists, however, did a controlled experiment to prove that DDT causes eggshell thinning, concisely summarized by Thomas Dunlap:

> They used American sparrowhawks [kestrels], birds that were closely related to the peregrine but could be raised in captivity. Comparing the

> eggs of birds fed small amounts of DDT and dieldrin (another insecticide implicated in reproductive failure) with eggs from a control group on an insecticide-free diet, they showed that both chemicals caused thinning of shells and lowered reproductive success.[6]

Beginning with the publication of *Silent Spring*, the debate on the environmental impact of DDT and its effect on human health—it had become widely known that DDT causes cancer in mice—made headlines in the newspapers and went on for years. The opponents of DDT tended to overdramatize their case and to appeal to human fears. But their central argument was correct and, finally, after long court battles and contentious hearings before the Environmental Protection Agency, they won their case. The debate was sometimes acrimonious and highly misleading, as illustrated by a passage in Paul and Anne Ehrlich's *Population, Resources, Environment*:

> The industry's tactics were typified by an editorial that appeared in the journal *Farm Chemicals* (January 1968), which not only labeled every biologist critical of current pesticide practice as a member of a "cult" and a "professional agitator," but also claimed that "scientists themselves literally ostracized Rachel Carson, and they will come to grips with this eroding force within their own ranks. Of course, it is not unusual that the character of the scientific community is changing. It may be a sign of the times. The age of opportunism!" This editorial appeared four years after Rachel Carson's death, yet the ghost of this remarkably sensitive and extremely capable marine biologist apparently still haunted those whose products she had found were dangerously polluting the environment.[7]

Nevertheless, in 1972 DDT was banned from use in the United States and was previously or has since been banned by most developed countries. Related insecticides that are as unsafe or even more unsafe than DDT have also been banned: dieldrin, aldrin, endrin, and chlordane. However, DDT and these related insecticides are still being used in some third world countries.

Even today, there are probably very few animals of any kind that do not carry at least traces of DDT in the fat of their bodies. This toxin and DDE are virtually everywhere. They move around the world on tiny dust particles carried by the wind. Every year during the heyday of its use, thousands of pounds of DDT landed on and were frozen in the Greenland ice cap, where no insecticide had ever been used. Although DDT is only slightly soluble in water, in the aggregate large amounts are transported in flowing water. It is also moved from place to place in the bodies of animals. By those various means, DDT has really gotten around. It has been found in salmon in the Great Lakes, in tuna fish caught far out to sea, in wild deer, and even in penguins on the Ross Ice Shelf in the Antarctic. It is in the bodies of people around the world and is secreted in mother's milk.

Thanks to the ban, the residue of DDT in the body fat and milk of humans has declined steeply. Bird populations have recovered. Peregrines, completely extirpated from the eastern states, have been reintroduced and are doing well—even nesting on the ledges of skyscrapers in cities such as New York, Chicago, and Toronto. Bald eagles, once almost gone, are back in full force. In winter, scores of them perch on trees along major rivers such as the Mississippi and the Illinois. Our southern seacoasts are once again graced by brown pelicans. In the 1960s and 1970s I was lucky to see even one American kestrel in a day as it perched on a wire as it hunted mice during the Illinois winter. Today I see one perched every few miles along country roads.

In his history of the DDT controversy, Thomas Dunlap wrote, "*Silent Spring* was at once an exposé of the damage pesticides were doing and might do to man and the environment, a report on less harmful methods of insect control, and a plea for a changed attitude toward nature."[8] "It is not my contention," wrote Carson, "that chemical insecticides must never be used. I do contend that we have put poisonous and biologically potent chemicals indiscriminately into the hands of persons largely or wholly ignorant of their potentials for harm."[9]

Carson's words conjure up the environmentally and economically realistic concept of insect control defined by Mary Flint and Robert van den Bosch:

> Integrated Pest Management (IPM) is an ecologically based pest control strategy that relies heavily on natural mortality factors such as natural enemies and weather and seeks out control tactics that disrupt these factors as little as possible. IPM uses pesticides, but only after systematic monitoring of pest populations and natural control factors indicates a need. Ideally, an integrated pest management program considers all available pest control actions, including no action, and evaluates the potential interaction among various control tactics, cultural practices, weather, other pests, and the crop to be protected. Under IPM, natural enemies, cultural practices, resistant crop and livestock varieties, microbial agents, genetic manipulation, messenger chemicals (such as sex attractants), and pesticides become mutually augmentative instead of individually exclusive or even antagonistic—as has been so often the case under pesticide-dominated control.[10]

In other words, an insect is best controlled by choosing judiciously from the arsenal of the numerous tactics and weapons available. What is now known as IPM, Integrated Pest Management, was the usual approach to insect control of entomologists of the old school. It was largely abandoned when DDT appeared on the scene, but was not long after that resurrected because of the dangers of modern pesticides to people and the environment and because the usefulness of insecticides is threatened because insects quickly become resistant to them when they are overused.

By 1938 a bacterial disease of the Japanese beetle and other "white grubs" that live in the soil was markedly reducing populations of Japanese beetles in the East Coast areas first infested. The spores of the causal bacterium, harmless to plants, humans, and other warm-blooded animals, can lie quiescent in the soil for years, germinating and causing disease, always fatal, only after they have been swallowed by a "white grub." This malady is known as the milky disease because infected larvae are opaque and milky white, because their blood is jam-packed with bacteria.

In 1955 a US Department of Agriculture Farmers' Bulletin by Walter Fleming reported the successful use of the milky disease to control Japanese beetles:

> The milky disease occurred only in limited areas in New Jersey and neighboring States when it was first found and studied. In such localities this disease appeared to be an important factor in bringing about a marked reduction in Japanese beetle numbers. . . . A program . . . using the milky disease organism in a practical way to reduce Japanese beetle populations was carried on by the Department of Agriculture in cooperation with State and other Federal agencies during the period 1939–53. By the time this program was concluded in 1953, almost 137,000 sites had been treated with the milky disease organism in 220 counties in 14 States and the District of Columbia.
>
> Under favorable conditions, the milky disease bacterium multiplies rapidly, but sometimes it may take a few years to control a well-established infestation of Japanese beetles. Nevertheless, significant effects have already been seen in the areas that were treated earliest.[11]

Rachel Carson posed a pregnant question. Why was this successful and environmentally friendly method not tried before Iroquois County was doused with dieldrin?

As Rachel Carson said, there are effective ways of controlling pest insects other than applying insecticides. For example, a major change in the midwestern cropping system benefited farmers by giving them a more profitable crop, soybeans, and has all but eliminated the sap-sucking chinch bug as a pest of corn. This change, planting soybeans rather than wheat as the alternative to corn, is a striking example of how a pest insect can be controlled by manipulating an agricultural ecosystem. Before this change, while chinch bugs were still serious pests of corn and before the modern synthetic insecticides were available, a corn crop could be protected from an invasion of chinch bugs only by physically trapping them in a plow furrow, an archaic, laborious, and tedious method.

17 THEIR PASSING FROM THE AGRICULTURAL SCENE

Chinch Bug

On a warm and sunny day in June 1928, a farmer and his horse were working hard to make a physical barrier between one of his wheat fields and an adjacent field of corn. He would rather have been angling for catfish in the nearby Sangamon River, but his livelihood was threatened by a horde of destructive, sap-sucking chinch bugs (*Blissus leucopterus*) that were about to march on foot from the drying, ripening plants in his wheat field to the nearby corn, which was still green, juicy, and therefore very attractive to the hungry bugs. But if the army of chinch bugs, then nearly full-grown and at their hungriest, had gotten into his corn, they would have sucked the plants dry, causing many to wilt and die. The catfish could wait, but the chinch bugs could not. Blocking them from invading his cornfield occupied the farmer for the next two weeks or even longer. I will come back to the fascinatingly archaic business of making a chinch bug barrier, but first we have to understand the habits and life history of this little bug.

The chinch bug, a native of the United States, feeds only on plants of the grass family, such as many wild grasses, wheat, other small grains, corn, sorghum, and milo. In the corn belt it was originally a denizen of the wild prairies, where grasses of several species were the dominant plants. Even today, in the heavily cultivated Midwest where virtually all of the prairie is gone, the chinch bug betrays its affinity for native prairie grasses by its choice of overwintering sites, which are preferably clumps of native bunch grasses that still grow here and there in fence rows and other bits of uncultivated ground. When Illinois and most of the other midwestern states were settled in the nineteenth century, the farmers unwittingly accommodated chinch bugs by planting mainly corn and small grains, especially wheat. This agroecosystem, with its plentiful growth of two cultivated grasses that replaced the native prairie grasses, was an ideal habitat for chinch bugs and persisted well into the twentieth century.

The small black and white chinch bugs survive the winter as diapausing adults tucked away in protected places, near ground level in the crown of bunch grasses when they are available or otherwise mostly under plant debris on the ground in fence rows, ditch banks, roadsides, or the edges of woodlots. Overwintering chinch bugs tend to be gregarious. As many as 5,000 of them per square foot have been found in suitable sites. In the central United States, they fly to these winter shelters in September and October and remain there until early spring, when they begin to fly off in search of their host plants.

When winter wheat was still a major crop in the corn belt, chinch bugs did not have to look far to find a spring host. Small green winter wheat plants, which had been planted and germinated the previous fall, survived the winter—even lethally low temperatures if they were covered by an insulating layer of snow—and were available to the bugs no matter how early they left their winter quarters. Once in a wheat field, the bugs fed, mated, and laid about 200 eggs per female at the rate of about a half dozen a day for three or four weeks. Chinch bugs place their eggs behind the leaf sheath, the lower part of the leaf that grasps the stem, or on the roots if the soil is loose. After the eggs hatch, the immature bugs, referred to as nymphs, feed and grow to adulthood in 30 to 40 days. Most of the nymphs do not mature until after the chinch bugs have left the mature drying wheat in June and

Chinch bugs on a cornstalk, an adult and on its left a young nymph and a nymph ready to molt to the adult stage.

moved to their summer host plants, a field of corn or occasionally to some other grass that is still green and luscious. Because the nymphs do not have fully developed wings, the bugs leave the ripening wheat and search for an alternate host plant on foot.

Our farmer's wheat had, of course, suffered some damage, but there was little that could have been done about that, and the damage was minimal because the still-growing bugs do not become maximally ravenous until after they leave the wheat field. In the old wheat/corn agroecosystem, the alternate host, the summer host, was almost always corn, which was, as it is today, abundant and was easily found by the bugs, especially if a field of it was planted next to a wheat field, as was often the case. After they reach maturity, the migrant bugs lay eggs, which give rise to another generation, which will grow to the adult stage as they feed on corn. It is these individuals of the second generation that will go into diapause and retreat to the protected sites where they will overwinter.

Precise directions for the arduous and time-consuming business of preparing and maintaining a chinch bug barrier appeared in 1914 in a pamphlet written for farmers by F. A. Fenton and C. F. Stiles, entomologists with the US Department of Agriculture:

1. Plow a furrow between the small grains and corn or sorghum, throwing the dirt towards the latter crops. The ridge should be 6 to 8 inches high with a flat surface at least 2 inches wide near the top as a base for the creosote line.
2. Thoroughly pulverize, smooth, and pack the soil on the furrow slice or ridge side. This may be done by dragging a straight log back and forth in the furrow; or the furrow can be smoothed down with the back of a spade.
3. Dig postholes in the bottom of the furrow or partly in the sloping side. These should be 1 to 4 rods apart [A rod is 16.5 feet.] and 18 to 20 inches deep. The number of holes will depend upon the number of migrating bugs; the more numerous the bugs the more holes are

needed. The tops of the holes should be flared and kept dusty to increase their efficiency in trapping the insects.[1]

A repellent such as tar or creosote was dribbled below the top of the ridge on the side facing the advancing chinch bugs and was renewed every 24 hours. The bugs, repelled by the creosote, were trapped in the ditch and, as they scurried about, many fell into the postholes. As least once a day, they were killed by plunging a post down into the posthole. Since chinch bug migrations were long and drawn out, the barriers had to be tended for two or three weeks to repair damage, renew the creosote barrier, and crush the bugs that had fallen into the postholes.

If well made and well tended, the barriers were effective. They certainly caught a lot of chinch bugs. C. M. Packard and his coauthors wrote in a 1937 *Farmer's Bulletin*, "In one instance 9 bushels of bugs were caught along 1/2 mile of creosote barrier in a week, and approximately the same quantity in the same barrier the next week. It was estimated that at least 60 million bugs were caught along this line in a week."[2]

Farmers have not made physical chinch bug barriers for over 50 years. After the development of the modern synthetic insecticides, they saved an enormous amount of time and effort by using narrow chemical barriers a few yards wide rather than physical barriers, an admirably parsimonious use of an insecticide. The insecticide must be one that poisons on contact, that need not be ingested by the insect because it is lethal if absorbed through the feet or some other part of the body. An anonymous publication of the US Department of Agriculture explains how to apply a chemical barrier by spraying a strip of dieldrin about 20 feet wide between the wheat and the corn.[3] The spray should be applied a few days before the bugs begin to migrate when the wheat plants begin to dry up.

The dieldrin barrier was effective for from 7 to 14 days—long enough to stop a moderately heavy invasion, but repeated spraying was sometimes necessary to cope with an exceptionally heavy migration or to renew a barrier destroyed by rain. Dieldrin and related chlorinated hydrocarbon insecticides, menacing environmental pollutants, have been banned from use in the United States, beginning with DDT in 1972. But that has had little or no effect on the control of chinch bugs because by then they were very seldom a serious agricultural problem since a new crop largely replaced wheat in the corn belt. This altered the ecology of the agroecosystem so that it was no longer friendly to chinch bugs.

In May 1923 W. L. Burlison and W. P. Flint of the University of Illinois Agricultural Experiment Station published a circular entitled *Fight the Chinch-Bug with Crops*. They pointed out that farmers could avoid chinch bug injury by planting crops other than small grains, crops that are not grasses and that chinch bugs will not attack, namely soybeans, cowpeas, fodder beets, buckwheat, sunflowers, or rape. They reasoned as follows:

> The chinch-bug is a grass-feeding insect; it has never been known to cause damage to any crop that does not belong to the grass family. Corn, as well as all of our small grains, belongs to the grass family, and therefore is subject to chinch-bug depredation. . . . There are, however, many crops which are not grasses that can be grown profitably in sections of Illinois now infested with chinch-bugs. By means of raising these crops each farmer can independently protect himself from chinch-bug injury, for all these crops are just as free from chinch-bug injury when grown on one farm in a heavily infested area as when grown on all the farms in the community.[4]

Above all, they recommended soybeans, which were then beginning to be popular with Illinois farmers as an alternative crop that is not attacked by chinch bugs. They did not, however, go so far as to recommend that all farmers substitute soybeans for wheat so as to end the wheat/corn feeding cycle of chinch bugs and thereby eliminate them as serious agricultural pests. Farmers, famous for their independence, would probably not have followed such a dictate anyway. But fortunately for corn belt agriculture and to the detriment of chinch bugs, it happened just the same, because in many areas soybeans are superior to wheat as a crop. They grow well in corn country and, averaged over a period of years, the yield from an acre of soybeans is worth more than the yield from an acre of wheat. Furthermore, soybeans are legumes, members of the pea family, whose roots, unlike those of wheat, corn, and most other plants, bear characteristic nodules that house bacteria which "fix" nitrogen, capturing it from the atmosphere and thereby fertilizing the soil with this major nutrient required by all crop plants and, for that matter, by all living things.

The most obvious and important effect of this profound change in the ecology of midwestern farm lands was, of course, the huge decrease in chinch bug populations, which freed the corn crop from one of its most destructive pests. There were certainly many other changes in the species composition of this agroecosystem, some subtle and not readily apparent—for example, organisms such as pathogenic fungi and parasitic insects that depend upon chinch bugs became less abundant. But other changes are apparent. Insects that feed on soybeans but not on wheat invaded the corn belt agroecosystem. The Corn/Soybeans Study Team of the National Research Council reported that the major insect pests of soybeans, an Asian plant not native to North America, are all native North American insects that have adapted to exploit this new plant: the green stink bug, Mexican bean beetle, and bean leaf beetle.

The green stink bug, named for the fetid odor of a repellent fluid it secretes, is a polyphagous insect that feeds on plants of several different families, among them cotton, several legumes other than soybeans, and several kinds of wild and cultivated fruits. When in April and May the adults leave their overwintering sites under plant debris in woods and fence rows, they then, according to Robert L. Metcalf and Robert A. Metcalf,[5] subsist by piercing and sucking juice from the fruits of various plants, including dogwood and elderberries. They do not invade soybean fields until August, when the still green and soft beans are developing in the newly formed pods. They lay large numbers of eggs on the soybean leaves, producing the one and only generation of the year. Adult and nymphal stink bugs pierce the pods to puncture and feed on the seeds, thereby killing them or causing them to be malformed. This damage is sometimes incorrectly ascribed to causes other than stink bugs.

When I spent a sabbatical leave doing research on the pest insects of soybeans at an agricultural research station of the Instituto Columbiano Agropecuaria near Palmira in the Cauca Valley of Colombia, I noticed soybean seeds that looked like they had been damaged by stink bugs. But the scientists at the station, most of them not entomologists, insisted that the malformation of the seeds was caused by an abnormality of the growth of the plant itself. When I told them it was caused by the feeding of stink bugs, they challenged me to prove it. A simple experiment—covering soybean plants, which had been freed of insects, with mesh cages that excluded insects and then putting stink bugs in some of them but not in others—proved that the malformation of the seeds was indeed caused by two tropical relatives of the green stink bug. As I reported in *Anais Da Sociedade Entomológica Do Brasil*,[6] over 76 percent of the seeds of soybean plants that had been caged with stink bugs had been punctured by them and were malformed, while only about 2 percent of those caged without stink bugs had been punctured. These latter few had been briefly exposed to stink bugs when the plants were just beginning to form pods and before they had been caged.

The Mexican bean beetle and two of its close relatives are the only members of the ladybird family that feed on plants. All of the others are predators that eat aphids and other small insects. This aberrant member of its family looks like a large, yellow ladybird beetle with small black spots. It feeds almost exclusively on legumes. Today its major host plant in some areas of the country is the soybean, but originally it fed on native legumes such as the beggar tick, whose tiny, triangular seedpods cling to your clothing, and cultivated ones such as lima beans and all kinds of snapbeans, which were grown by Native Americans thousands of years before the first Europeans found their way to the New World. (The original inhabitants of the New World contributed many other food plants to the European diet: corn, potatoes, sweet potatoes, tomatoes, peppers, chocolate, and avocados, to mention just a few.)

The Mexican bean beetle is now in the process of adapting to a new leguminous plant, the exotic soybean. It currently infests this crop here and there in the United States but by no means everywhere. The adults, which overwinter on the ground under plant debris, straggle out of their winter quarters for almost two months in spring and early summer. They move to their host plants and lay large masses of yellow eggs on the underside of leaves. Both larvae and adults feed from the underside of the leaves, partially "skeletonizing" them by eating everything except the veins and the upper epidermis. The lacelike appearance of these leaves is characteristic of Mexican bean beetles.

The bean leaf beetle is another specialist that feeds only on legumes—today mainly on soybeans, but before the appearance of this plant in North America, on the same wild and cultivated native legumes that Mexican bean beetles utilize. The adults, only about a quarter of an inch long, are highly variable in color and markings, but are typically reddish to yellowish and have black markings. They chew large more or less round holes in leaves. Although it looks severe, this damage and leaf feeding by other insects will not necessarily decrease yield, because the plants can tolerate the loss of a surprisingly extensive area of leaf surface without a loss of yield: 20 percent when they are actually developing seeds and as much as 30 percent at other times. The plants "budget" for the possible loss of leaf area by growing more leaves than they need to produce their optimal quota of offspring, which are, of course, the seeds the farmer harvests. Bean leaf beetle larvae live in the soil at the base of a leguminous plant and feed on the roots and root nodules that house the bacteria that "fix" atmospheric nitrogen. Little is known about the effect of larval feeding on the plant, although this kind of injury is likely to have a serious impact.

Bean leaf beetles overwinter as adults, mainly in leaf litter and other debris on the ground in places such as fence rows, field margins, and woodlands. They become active in April, well before soybean plants are available to them; and until they are, the beetles feed on wild legumes and, in a pinch, even make do with some plants that are not legumes. As soon as soybean plants break the surface of the soil, the bean leaf beetles move into the fields where they feed and lay their eggs in the soil at the base of a plant. These eggs, the first generation of the season, mature to become the parents of a second generation, which in autumn retreats to overwintering sites.

Extensive surveys made in Missouri by Norman Marston and colleagues[7] showed that the soybean ecosystem is also inhabited by many other insects and spiders, at least 175 species. Among the other insects that feed on soybean plants, usually less destructive than stink bugs, bean leaf beetles, and Mexican bean beetles, are wireworms, the root-feeding larvae of click beetles; the foliage devouring Japanese beetle; the red-legged and differential grasshoppers, which feed on foliage in both their nymphal and adult stages; and the leaf-feeding caterpillars of the beautiful painted lady butterfly and of two moths, the green cloverworm and the garden webworm. In addition they found 19 other aphids, leafhoppers, thrips, flies, beetles, and a moth that also feed on soybean plants. There were 10 parasites and predators of the soybean feeders, and a miscellany of over 140 other less abundant flies and beetles that are probably important as scavengers and as an alternate food source for predators.

Substituting soybeans for wheat, as we have seen, had profound ecological effects on the midwestern cropping system. Chinch bugs all but disappeared, and a suite of once-uncommon and innocuous native insects, finding a new and abundant food resource in soybeans, burgeoned in numbers and some became immensely important in the agroecosystem. There have been other cases of native insects adapting to foreign crop plants that agriculture brought to them, but the most convoluted and interesting of them is the saga of the coming together of the common white potato (which originated in the Andes Mountains of Bolivia and Peru) with a then all-but-unknown beetle of the Great Plains of the United States.

The potato's circuitous route from the Andes to the American West, so interestingly described in 268 readable and fascinating pages by Larry Zuckerman,[8] twice crossed the Atlantic and was long in both distance and duration. The "humble spud," as Zuckerman calls this bounteous plant, was first cultivated about 7,000 years ago on the high altiplano of Bolivia. Long before the Spaniards came to enrich themselves with gold at the expense of the Incas, the potato had become the staple of the Andean diet. The conquistadors brought the potato to Spain in 1570, but it was slow to be accepted because it was a relative of the deadly nightshade and because it was thought of as a lowly food, the food of the conquered Indians. But by the late eighteenth century it was an important crop in most European countries, especially in Ireland, where there was massive starvation in the 1840s when the blight, a fungal disease, destroyed the potato crop. The early settlers brought the potato back across the Atlantic to the North American colonies, and as the pioneers moved west across the continent, they planted potatoes wherever they settled. In the 1850s they reached the Great Plains.

There the potato was greeted by a beetle, now known as the Colorado potato beetle, which was to become the worst insect enemy of this important crop. Just as the Spaniards transported the potato from the Andes to Europe, they were responsible for this beetle's invasion of

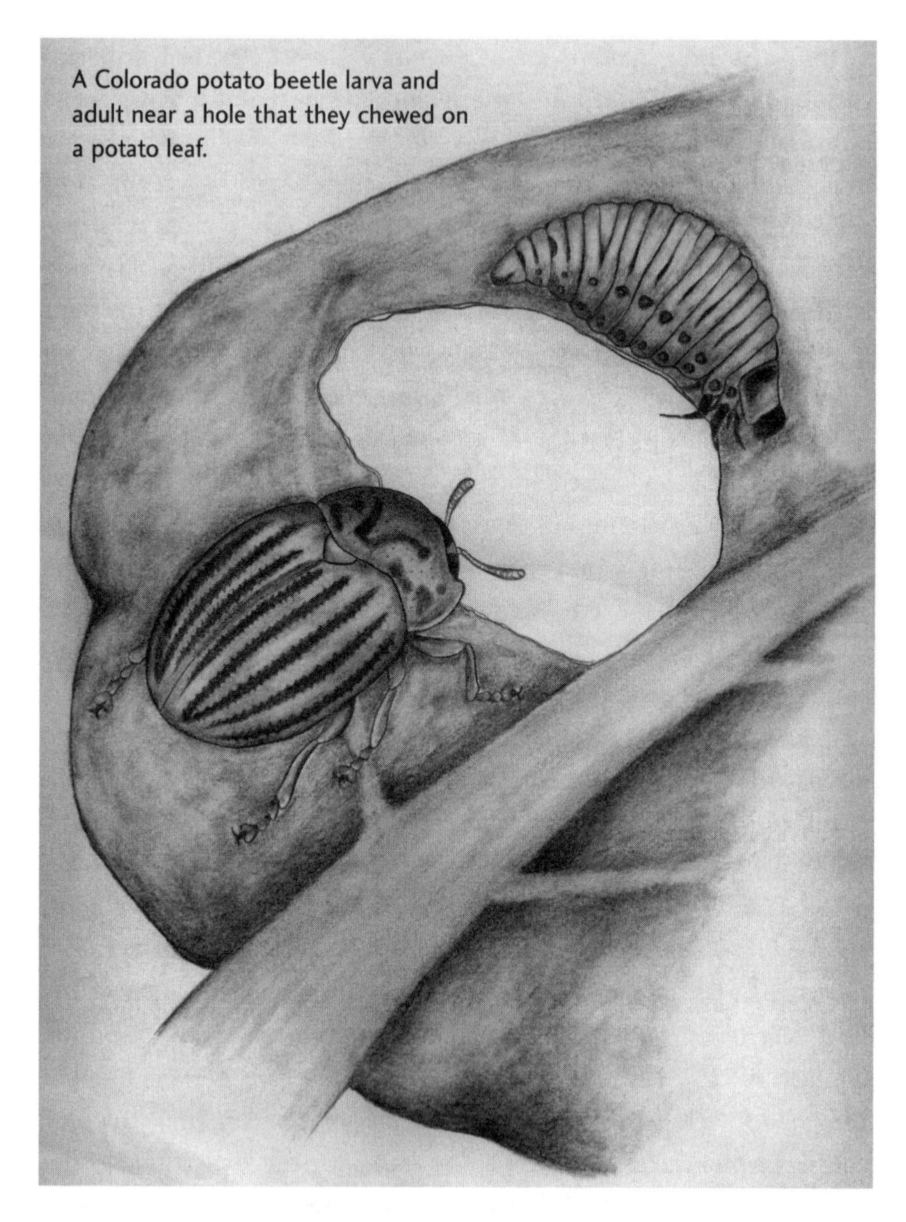

A Colorado potato beetle larva and adult near a hole that they chewed on a potato leaf.

the Great Plains, a story engagingly told by Wenhua Lu and James Lazell.[9] The beetle and its host plant, the buffalo bur (*Solanum rostratum*), lived and still live in central Mexico, where the buffalo bur's close relative, the potato (*Solanum tuberosum*), did not grow. The

Spaniards brought domesticated cattle to Mexico in the sixteenth century, and by about 1680 Mexican vaqueros were driving them, many with burry seedpods clinging to their fur, to market in Texas, where the buffalo bur established itself, as it had all along the route. From there its seeds were carried north to the Great Plains, clinging to the coats of migrating bison. There was now a highway of buffalo bur plants along which the beetles moved from Mexico to the Great Plains.

In time, Colorado potato beetles left the buffalo bur to feed on the leaves of potatoes planted by the new settlers. Once this insect had adapted to the potato as a food plant, it was possible for it to do something that had previously been impossible for it. It migrated eastward along a broad highway of potato patches, moving from patch to patch at an average rate of about 85 miles per year. It reached Illinois in 1864, Ohio in 1869, and the Atlantic coast in 1874. Shortly after World War I it appeared in Europe, soon spread throughout that continent, and continues to move eastward across Eurasia wherever potatoes are grown.

Just as the agroecosystem of the corn belt was significantly changed by substituting soybeans for wheat, the presence or absence of just one species of insect can have an enormous impact on an ecosystem. St. John's wort, known as Klamath weed in California, is an invader from Europe that by 1945 virtually covered almost 8,000 square miles of rangeland in the northwestern United States, having all but vanquished the native grasses and other forage plants. A new ecosystem, not favorable for grazing, had been established. But all this was changed by just one insect, a leaf beetle, *Chrysolina*, that feeds on Klamath weed and had been introduced from Europe as a biological control of the weed. Within a decade Klamath weed was controlled by *Chrysolina* and had virtually vanished from the range.

An ecological survey of this area done today would find Klamath weed growing mainly in the shade and supporting only sparse populations of *Chrysolina* beetles. If we did not know about the biological control of this plant, we might falsely conclude that it is shade loving and doesn't grow well in the sun, and that *Chrysolina* is naturally rare. But we know Klamath weed grows all too well on the sunny open range, and that the beetle was obviously far from rare when it was destroying these plants on open grazing lands. The truth is that today this plant rarely survives in the sun because the beetle, which thrives in the sun, soon kills it. Shady places are refuges for St. John's wort, because the beetles do not thrive there and consequently do not become numerous enough to kill their host plants. These facts led Peter Price, in his book on insect ecology, to ask two penetrating questions:[10] "Could the ecology of these species be properly understood if this interaction had not been witnessed?" The second question follows from the first. "How many other plant species have distributions dictated by herbivore pressure?" The main point is that, unknown to us, all over the world plant-feeding insects are probably limiting the population expansion of many plants by driving them out of the habitats where they grow best.

The Hessian fly, which we will consider next, can be controlled by a far less drastic change in the agricultural system. All that must be done is to sow winter wheat in the autumn when or after Hessian fly females have died. If there are no females to lay eggs on young wheat plants, there will be no infestation. The "fly-free date" could not be determined if we did not have a thorough understanding of the life history and behavior of this insect, one of the most destructive pests of wheat.

18 SYNCHRONY WITH THE SEASONS

Hessian Fly

As an American farmer harvested his wheat in the summer of 1779, during the War for Independence, he saw that his crop had been severely damaged. Walking through the field swinging his scythe, he discovered that many plants had been killed; some were stunted and had not sprouted the usual tall stalk topped with a head of grain; and many of the stalks that had formed a head were broken over. In his area, the then rural western end of Long Island—later to become New York City's boroughs of Brooklyn and Queens—other farmers had the same problem, which until then no American farmer had ever seen. They soon found that the damage was done by a small, legless, white maggot only 3/16 of an inch long. But where had this pest come from and how had it gotten to their fields? As a matter of fact, three years earlier, a British force including 8,000 German mercenaries, most from Hesse, had invaded Long Island. The Hessians had undoubtedly dumped the matted European wheat straw from their bedding and replaced it with fresh straw. That old straw was the probable source of the offending insect, a new invader of North America soon dubbed the Hessian fly (*Mayetiola destructor*). Today this insect is found wherever winter wheat is grown in North America, and now these tiny insects number among the most destructive enemies of wheat.

Winter wheat, unlike most other crops, is sown in fall rather than in spring and is harvested in early summer. The seeds sprout in fall and, after growing to the height of a few inches, become dormant for the winter, surviving even potentially lethal low temperatures if they are covered with an insulting blanket of snow. (If it gets very cold in the Dakotas when there is no snow cover, the market price of wheat goes up.) In spring, the plants resume growing, and in July, dry and golden ripe, they are ready to be harvested. As you will see, the life history and habits of the Hessian fly predispose it to exploit winter wheat as a food plant.

Hessian flies, explained W. B. Cartwright and E. T. Jones,[1] have two generations a year: adults are present in autumn and again in spring, in central Illinois in September and April. The adults do not feed, rarely survive for more than three or four days, and cannot fly for long distances under their own power. The winds, however, can carry them for miles, which is the only way they can disperse over a long distance unless they are transported in bales of straw. On average, mated females lay from 250 to 300 eggs on the lower leaves of their host plant, which is always a member of the grass family—in North America mainly wheat, occasionally rye or barley, and only rarely a few wild grasses. An egg is only about 1/50 inch long, glossy red, slender, and almost cylindrical, with both ends bluntly rounded. The mother lays them end-to-end in one of the lengthwise grooves on the upper side of a leaf, often in strings of about a dozen. When seen through a magnifying glass, they look like a miniature string of hotdogs.

After anywhere from 3 to 10 days, depending upon the prevailing temperature, the eggs hatch, and the tiny maggots crawl down the groove and settle behind the leaf sheath. There they feed by rasping on the stem and drinking the sap that oozes from the wound. They do not burrow into the plant, but their rasping, which can be likened to open-pit mining, excavates a depression into the stem. In fall, when the wheat plants are small, this injury may kill a plant outright or weaken it so much that it cannot survive the winter. Infested plants that do manage to survive are stunted and few produce a head of grain. Maggots of the spring generation seldom kill a plant but their feeding may so weaken the central stalk that it breaks over when the head of grain becomes ripe and heavy.

After about two weeks, the maggots are full-grown. They stay within their pit and form a puparium in the same way that a house fly does. The brown puparium, which resembles a flax seed, is often referred to as the "flax seed stage." The fall generation overwinters as a maggot in the puparium. In April the maggot pupates and soon emerges as an adult. The maggots of the spring generation form the puparium in June but do not become adults until September, just in time to lay their eggs on seedling winter wheat plants. After the wheat is cut in early July, many puparia survive in the stubble left standing in the field. Even if a farmer double-crops, planting short-season soybeans after the wheat harvest, the puparia survive because the stubble is not plowed under before the soybeans are planted.

Once Hessian flies have infested a field, nothing can be done to control them. Insecticides are useless because the maggot, in its snug retreat behind a leaf sheath, is safely shielded from harm. The only recourse is to prevent an infestation from becoming established in the first place. Varieties of wheat resistant to the Hessian fly have been developed and at first they greatly reduced the incidence of damage. But natural selection and evolution soon changed that. Races of the Hessian fly that can survive and thrive on one or more of the resistant varieties appeared, so the resistant varieties gave less protection than they did at first. Happily, a sound knowledge of the Hessian fly's natural history revealed an elegant and infallible way to prevent the fall brood from infesting a planting of winter wheat.

The solution is to sow the field late enough in the fall so that the plants will not come up until after the last adult flies have laid their eggs and died. This is possible because the flies emerge from their puparia more or less in synchrony and, since they do not feed, they live for only a few days. Fortunately, all the flies are dead before it is too late to plant winter wheat. The "fly-free date," the safe date on which to plant wheat, differs with the latitude. It is, of course, earlier in the north, where the growing season is short, and later in the south with its longer growing season. Illinois, almost 400 miles long from its northern border with Wisconsin to its southernmost tip at the town of Cairo, is divided by entomologists into 13 zones with different fly-free dates—September 18 near the Wisconsin border, September 30 in the central part of the state, and October 12 in the vicinity of Cairo.[2]

The usefulness of the fly-free date in controlling the ravages of the Hessian fly became apparent only after people became aware that the seasonal cycles of the fly and of winter wheat are synchronized. In other words, they had come to understand the phenology of the plant and the insect. The phenology of an organism is the timing of the seasonal pattern of its development and of its feeding, mating, egg laying, and other behaviors as they relate to the climate and especially to the development and behaviors of other organisms. Timing is all-pervasive in nature. As Ecclesiastes 3:1–8 tells us: "For everything there is a season, and a time for every purpose under heaven: a time to be born, and a time to die; a time to plant and time to pluck up that which is planted."

The lives of the insects—of all organisms in an ecosystem—are four-dimensional. In other words, the position of organisms in three-dimensional space and their position in time—the fourth dimension—are both important to their own well-being and to the continued survival of the ecosystem in which they live and upon which they depend for their very existence. The propitious time for the many activities of an insect are dictated not only by the climate, but ultimately by the availability of such resources as mates, food, shelter, and the need to escape predators and other enemies. Seasonal factors such as day length and temperature are cues that prompt insects to terminate diapause, mate, lay eggs, or otherwise act at the time most favorable for members of their species. In this way the rhythm of an insect's life is (shown below) coordinated with the life rhythms of many other organisms. These phenological relationships facilitate the coexistence and integration of the great diversity of plants, insects, and other organisms in an ecosystem.

The phenology of the native North American bean leaf beetle and its host plants—all legumes, members of the pea family—offers a particularly dramatic example of the importance of timing. When the Asiatic soybean was planted in the Midwest, it soon became the most abundant of the bean leaf beetle's host plants. Fed by soybeans, which were everywhere and enormously abundant, the population of this insect soared and became much too large to be supported by just the miniscule surviving population of its native host plants, such as the beggar tick, the hog peanut, and the common garden bean, a native legume first cultivated by Native Americans.

Adult bean leaf beetles leave their winter quarters in fence rows and woodlots in April, but soybeans are planted in May and the new shoots emerge from the soil in late May or early June. The beetles must wait a month or more to feed on soybean seedlings and to lay their eggs in the soil at the base of these plants. In the meantime, they survive by feeding in fields of the leguminous clovers and alfalfa, and Charles Helm and his colleagues[3] discovered that they will even feed on some nonlegumes such as nettles and a wild *Euonymous*. But they lay no eggs in association with any of these plants, waiting until soybean seedlings emerge from the soil. The timing is close and critical. Marcos Kogan and I[4] saw just how critical it is when the unusually wet spring of 1974 greatly delayed the planting of soybeans in our area. The beetles had become active in April, as they usually do, but soybeans did not germinate until late June and early July, over two months later. In the absence of their new and usually superabundant host plant, the great majority of the beetles starved to death before they laid any eggs. Their population crashed and remained unusually low for the next two years. But a small population of bean leaf beetles did survive on the scarce native legumes.

Carol Augspurger's[5] elegant experiments in Panama demonstrated the importance of phenology by revealing the immense reproductive benefits a forest shrub gains by flowering in close synchrony with the other members of its species. These shrubs of the genus *Hybanthus* develop many buds in the wet season, but do not flower until about a week after a heavy rain that interrupts the drought of the following dry season. Augspurger, a plant ecologist now in the University of Illinois's Department of Plant Biology, induced some individual shrubs to flower out of synchrony with the other members of their species by creating an artificial rain. She watered them with a sprinkler for seven hours prior to the first dry season rain. Shrubs that flowered in synchrony with each other after a natural rain produced an average of 658 seeds each, while individual shrubs prompted to flower out of synchrony with the natural population by an artificial rain produced an average of only 62 seeds each. The evolutionary fitness, the reproductive success, of a flowering plant is, of course, measured by its production of viable mature seeds—its offspring.

How many seeds a *Hybanthus* shrub produces depends mainly upon how many of its flowers are pollinated by bees and other insects and, to a lesser extent, upon how many of its developing seeds are destroyed by insects that ecologists call "seed predators." Augspurger showed that 78 percent of the flowers of shrubs in synchrony with the main population had been pollinated, but only 40 percent of those on shrubs out of synchrony. Shrubs out of synchrony were less likely to be found by pollinators, and their developing seeds were more likely to be found and destroyed by seed predators than were those of shrubs in synchrony. She attributed this latter difference to what ecologists call "predator satiation," the prey, in this case *Hybanthus* seeds, being so numerous in season that the predators' appetites were sated well before they ate all the prey.

Just as palatable black swallowtails gain protection from insect-eating birds by mimicking toxic butterflies, other insects—mainly flies, notably flower flies—profit from mimicking venomous stinging models, mainly social wasps and bumblebees. The phenology of these mimics and their models is contrary to a "law of mimicry," that mimics are best protected from birds if they "get lost in the crowd" by occurring when their models are the most numerous. At least in North America, some mimetic flower flies—the most deceptive of them are "high fidelity" mimics that may fool even entomologists—are, as Joe Sheldon and I[6] discovered, present as adults only in spring when they outnumber their models rather than in midsummer when they would be greatly outnumbered by the then much more numerous models. Wasp and bumblebee colonies, unlike those of honeybees and ants, die off in autumn; the only survivors are mated queens that overwinter and found new colonies in the ensuing spring. Consequently, there are few models in spring, but because the colonies grow rapidly, the population of stinging models increases exponentially and is at its peak by midsummer.

The mimics benefit by coinciding with the spring flush of nectar plants and presumably with the moist conditions of spring that favor their larval offspring, which live in moist rotting wood or water-filled cavities in trees. But how is it possible for these mimics to survive in spring when they outnumber their models, and hordes of hungry insectivorous birds have returned from their southern winter quarters? The answer is that these birds remember painful encounters with wasps and bumble bees from the previous summer and, therefore, shun the mimics, which they mistake for real wasps and bumblebees. In midsummer, young birds learn to shun bumblebees and wasps very quickly, because when they leave their parents to hunt on their own, the stinging models are very numerous and there are no harmless mimics to dilute the learning experience. A number of investigators have found that several species of birds have surprisingly long memories. For example, Miriam Rothschild[7] showed that a captive crow rejected on sight noxious insects that it had not seen for 9 months, and other investigators found that some birds rejected noxious insects that they had not seen for as long as 15 months.

The mutually beneficial interaction of predaceous ants and North American black cherry trees is an extraordinary example of phenological synchronization, because the tree can terminate the interaction when the ants can no longer benefit it. The tree attracts the ants by secreting sweet nectar—a favorite food of the ants—from glands on its young leaves. The ants reciprocate by killing and eating insects that feed on the tree's foliage, the most destructive of which are by far eastern tent caterpillars, which live in large colonies housed in large, white, silken tents in a crotch of the tree. "On the first day of bud break," notes David Tilman,[8] "I observed several occasions in which a nectar-foraging [ant] encountered first instar [tent caterpillars], grabbed a larva in its mandibles, and carried it back to the [ant's] nest." When he returned to the site two days later, he found that all the tent caterpillar larvae in this colony were gone. Several similar observations he made during the next two weeks showed that the ants often destroyed whole colonies of the tiny first instar larvae. If their nest was nearby, the ants did a thorough job of ridding the tree of tent caterpillars. Tilman found that later in the season not one of 81 saplings within about 33 feet of an ant nest was still occupied by a tent caterpillar colony, but 52 of 267 saplings from 82 to 131 feet away from an ant nest were occupied by colonies of caterpillars which were large, voracious, and capable of completely defoliating a young tree. About three weeks after bud break, all the caterpillars on a tree had either been destroyed or had grown too large for the ants to handle. The ants are then no longer useful to the tree. At this point the tree stopped secreting nectar, which is costly to produce, and ants—which had been coming in droves—seldom came to the tree.

A food chain of trees, caterpillars, and birds persists because these organisms are more or less phenologically synchronized with each other. The chain starts in spring when, triggered by warmth, the buds on trees such as oaks and maples break open and new leaves, at first as tiny as a mouse's ear, begin to unfurl. As are almost all other organisms, an oak tree is beset by enemies, many of them caterpillars that feed on its leaves. When very young, the leaves are nutritious and poorly defended against insects. But as they mature, Paul Feeny[9] demonstrated, they become too tough for most insects to chew and their nutritional value declines precipitously as their protein content is halved and their content of tannins, which interfere with the digestion of proteins, increases. In England the pupae of winter moth caterpillars that fed on young leaves weighed on average 30 milligrams, while those that had been raised on old leaves weighed only 10 milligrams. As is to be expected, many insects, among them cankerworm caterpillars in North America, feed on young oak leaves in spring, but few insects feed on older leaves.

According to John Schneider,[10] many spring caterpillars are not always precisely synchronized with the opening of the buds because the rate of their development in the egg is affected by the variable temperatures of spring somewhat differently than is the development of the buds. Being out of synchrony with its host plant can be a disaster for a caterpillar or any other insect. If they are three or four days too early, they will starve to death. If they are too late, the leaves will become too mature for them before they finish growing. The caterpillars may not survive on old leaves, and even if they do, they will be stunted and lay less then the normal complement of eggs. But the fall cankerworm, one of the many species of caterpillars known as measuring worms, has evolved a way of synchronizing with both red maple, which begins to leaf out in mid-April, and black oak, which begins to leaf out as much as a week later.

There are at least two different clones of the fall cankerworm; the members of one clone hatch from the egg early so as to coincide with the leafing out of red maples, and those of the other hatch later so as to coincide with the leafing out of the black oak. To understand how fall cankerworms have managed this, we must look at it in the context of the fall cankerworm's life cycle. This insect overwinters in the egg stage on the bark of its host tree. The larvae that hatch from the eggs in spring feed for three or four weeks and then burrow into the soil near the host tree and molt to the pupal stage. Winged males and wingless, parthenogenetic females emerge in the fall. Odd as it may seem, and unlike almost all other parthenogenetic insects, the females will not lay eggs unless they mate with a male. However, the male's sperm do not fertilize the eggs; therefore, a females' progeny constitute a clone. Her female progeny are, thus, exact genetic copies of her, and their eggs will hatch at the time in spring characteristic of their clone—those of the red maple clone earlier than those of the black oak clone.

The northward migration of some insectivorous birds, such as bay-breasted, Cape May, blackpoll, and blackburnian warblers, is in most years more or less synchronized with the appearance of young leaves and the coincident abundance of caterpillars. In some years the migrating birds arrive in central Illinois in time to feast on the caterpillars, but because their northward progress is influenced by the vagaries of the weather, in other years they arrive too early or too late.

Many entomologists have said that insects are their own worst enemies. Virtually all species of insects are beset by other insects that parasitize them or prey on them. Parasites and/or predators are often the most effective factor controlling an insect population. This is dramatically confirmed by many instances of successful biological controls—the use of a parasitic or predaceous insect to control a pest insect. One of the first and most dramatically successful biological controls was the importation and establishment in California of an Australian ladybird beetle to control the cottony cushion scale, which was accidentally introduced from Australia and threatened to destroy all of the citrus orchards in California. Let us see how this amazing story unfurls.

19

AN INSECT TO CONTROL ANOTHER INSECT

Cottony Cushion Scale

On an October day in 1886, Charles Valentine Riley, who had 15 years earlier rescued the French wine industry from certain destruction by the grape phylloxera, addressed a convention of the California citrus growers' association. His subject was the insect scourge that was then well on its way to obliterating California's infant citrus industry as it spread like wildfire from one orange grove to another. This scourge was the cottony cushion scale, so called because for most of its life it is covered with white waxy threads and because the female forms a wax-covered, cushionlike egg mass that remains attached to its body. Paul DeBach,[1] an entomologist at the University of California at Riverside, observed that infested trees were so densely shrouded with egg masses that they looked as if they were covered with snow. A citrus grower recalled the catastrophe that had threatened his grove before Riley devised a brilliant strategy that stopped this scale insect in its tracks: "The white scales were incrusting our orange trees with a hideous leprosy. They spread with wonderful rapidity and would have made citrus growth on the whole North American continent impossible within a few years."[2]

An insecticide commercial on television features a confused individual who—hoping for a one-size-fits-all way to kill pest insects—says, "I don't want to know their names, I just want to get rid of them." But entomologists know better: because every species is unique, there are no panaceas, no cure-alls. Control programs must be tailor-made; and the more you know about a pest the more likely you are to get a "good fit." After the crucial first step—identifying the insect, learning its name—you can look up in a library or on the Internet what is already known about it. Once Riley and his associates identified the cottony cushion scale (*Icerya purchasi*), they learned that in its homeland, Australia, it is never abundant enough to be destructive. Building on this knowledge and the results of his own research, Riley,[3] as we will see, devised a self-perpetuating control that all but eliminated cottony cushion scales from California by 1889 and that has since kept them in check for well over a hundred years.

The cottony cushion scale, like most of the other 6,000 species of scale insects, has undergone what some biologists have called "retrogressive evolution." In other words, except for first instar nymphs and adult males, their anatomy has become simplified to accommodate their unusual lifestyle. Scale insects are often referred to as "plant parasites," because for most of their lives they are sessile, immobile, and fixed in place by their sucking mouthparts, which are permanently embedded in the tissues of the food plant. They are little more than a blob of a body—covered with waxen white threads as is the case with the cottony cushion scale, or, as in the armored scale insects, by a waxy scale that the insect secretes and that covers its entire body as a camper is "covered" by a pup tent.

As do most and probably all female scale insects, adult female cottony cushion scales broadcast a volatile sex attractant pheromone that attracts tiny, flying adult males, which have all the usual appendages except that their mouth parts are vestigial and cannot be used to suck sap. Mated females lay wax-covered masses, cushions, of as many as 1,000 bright red eggs. After anywhere from a few days to a few weeks, depending upon the temperature, nymphs with eyes, antennae, legs, and mouthparts hatch from the eggs. Agile, active, and known as crawlers, they seek a place to settle down. Only in this stage of the life cycle can scale insects colonize new areas. They then settle down in the place where, as nymphs of all ages and sexes, and as adult females, they will live for the rest of their lives. Next they undergo their first molt, and in the process lose all of their appendages except the mouth parts, and secrete the waxy threads that adorn their bodies. After feeding avidly and molting several times, they become adults and the life cycle begins again.

An entomologist at the Virginia Polytechnic Institution, Michael Kosztarab,[4] asked if scale insects are good for anything. He recognized that scales are notorious plant pests, but that historically they have been more commercially useful than has any other insect group of comparable size. Red dyes were made from adult females of several Eurasian scales but most notably from the cochineal scale of the New World, which feeds on prickly pear cactus. The Incas dyed their ceremonial robes with cochineal extract, and the Aztec emperor Montezuma demanded tax payments partly in dye grains (dried cochineal bodies) from his slaves. This insect is still cultivated in the Canary Islands and in Central and South America as the source of a commercial dye called carmine red and used mainly to color candy and other foods. Lac, the secretion of an Asian scale insect that encrusts the branches of trees in India and Burma, is used to make shellac and sealing wax, and is an ingredient in many other substances. The ancient lac industry, certainly several thousand years old, is still an important export business in Asia. Scale-insect secretions have had and still have many other uses, for example, in the Old World to make candles and by Native Americans to waterproof baskets and as chewing gum, delicious because the scale is covered with sweet honeydew.

As to the cottony cushion scale, none of the limited variety of insecticides then available gave adequate control of this devastating pest. But Riley came to the rescue. His own studies of the biology of this insect and, above all, the knowledge that, although it is overwhelmingly abundant in California, its population is low and economically insignificant in its native Australia, gave him an idea that was to bear wonderful fruit. He hypothesized that in Australia the cottony cushion scale is controlled by natural enemies—parasites and predators—that had not accompanied it when it was unintentionally brought to California. Riley proposed that this scale could be controlled in California by importing and establishing populations of insects that are its natural enemies in Australia.

Riley recommended this tactic, now known as biological control, to the California Citrus Growers' Association: "It has doubtless occurred to many of you that it would be very desirable to introduce from Australia such parasites as serve to keep this fluted scale in check in its native land. . . . This state—yes, even Los Angeles County—could well afford to appropriate a couple of thousand dollars for no other purpose than the sending of an expert to Australia to devote some months to the study of these parasites there and to their artificial introduction here."[5] This was accomplished using federal funds, but only after Riley overcame difficulties to be discussed shortly. The insect that controlled the cottony cushion scale was not a parasite, but a predaceous ladybird beetle, the red and black vedalia. Its introduction was an astounding success.

Riley wanted to send Albert Koebele, an immigrant from Germany and one of his most capable associates, to Australia to find and send back to California natural enemies of the cottony cushion scale. But he feared that the US Congress would not appropriate the necessary funds. Riley was right. Congress had put a rider prohibiting foreign travel by employees on the appropriation bill for the Department of Agriculture. It was aimed at Riley himself, whose frequent junkets to Europe were made at government expense. But Riley circumvented Congress's ban on foreign trips.

In an introduction to Koebele's report of his work in Australia, Riley wrote:

> Failing to secure a specific appropriation from Congress for this purpose . . . and failing also to secure the removal of the clause restricting travel to the limits of the United States, we were led to accomplish the result through the kindness of the Department of State, in connection with the Melbourne Exposition, an arrangement having been made whereby two of the salaried agents of the Division should be temporarily attached to the Commission, their expenses, outside the United States to be defrayed by the Commission, within the sum of $2,000.[6]

Koebele found cottony cushion scales scarce and difficult to find in Australia. But in a garden in Adelaide he found a few scales and vedalia beetles on an orange tree. "I discovered there," wrote Koebele, "for the first time, feeding upon a large female [cottony cushion scale], the Lady-bird, which will become famed in the United States—*Vedalia cardinalis*."[7] He found more vedalias elsewhere and in 1888 and 1889 shipped many to California. They were packed in wooden boxes with branches bearing scales for them to eat, and during the long journey across the Pacific Ocean the boxes were kept on ice. Some of the ladybirds died in route, but during the last two months of 1888 and January 1889, the 129 that survived were released on a caged tree in an orange grove in Los Angeles. Another 385, most of them brought by Koebele when he returned from Australia, were released in groves elsewhere.

The caged vedalias multiplied and by early April of 1889 had eaten almost all of the scales on the tree. When the cage was opened, the beetles moved to nearby trees. Within six months of their first introduction, the voracious vedalias had virtually wiped out the scales in the Los Angeles orchard and had spread to nearby orchards. Vedalias were sent to other growers, and as DeBach[8] reported, "By June 12, 2 months after the cage was opened, 1055 vedalias had been distributed to 208 different growers." By the end of 1889, the scale was no longer a threat anywhere in California. Ever since, populations of the scale too small to be a threat have been kept at this low level by correspondingly small populations of vedalias—except for a short time in 1946 and 1947 when the ill-advised use of DDT, which did not kill the scales, all but wiped out the vedalias. The problem was soon solved by discontinuing the use of DDT.

This was not the first biological control, but it was so spectacularly successful that it attracted attention worldwide. As Paul DeBach says, it "established the biological control method like a shot heard around the world."[9] Encouraged by California's resounding success, entomologists all over the world developed, and still develop, biological controls for pest insects and also for weedy plants.

In the South Pacific, caterpillars of the *Levuana* moth, a highly destructive pest of coconut palms, had killed virtually all of these trees on some islands of the Fiji archipelago and were well on the way to wiping out Fiji's well-established and important copra industry. (Copra, dried coconut meat, is the source of coconut oil.) Because it had no parasites in Fiji, J. D. Tothill and his colleagues[10] believed that *Levuana* is not native to this archipelago and must have come from elsewhere in the region. But a search of southern Asia, the East Indies, and South Pacific Islands other than Fiji turned up no *Levuanas* and, of course, no predators or parasites of this moth to introduce to Fiji as a biological control. But Tothill and his group took another tack. Because they knew that most parasites are adapted to utilize several related species, they brought to Fiji a tachinid fly, *Ptychomyia*, that parasitizes *Artona*, a moth related to *Levuana*. This was a good move. Three years after its introduction, *Ptychomyia* had so reduced the *Levuana* population that visiting entomologists could not find these moths without help.

According to Donald Naflus,[11] an entomologist at the University of Guam, a caterpillar that feeds on the buds, young leaves, flowers, and fruits of mango trees was accidentally introduced on the island of Guam some time before 1940. No one knows where these particular invaders, mango shoot moths, came from, but their species is widely distributed in tropical Asia and the Pacific Islands. It is seldom destructive in its native range—a minor pest at most—because it is held in check by parasites and other natural enemies. But few parasitic or predaceous insects attacked it on Guam, so the little mango shoot moths overran the island and mango trees produced little or no fruit.

In 1986–87 parasites from southern India were released on the island and became established, a tachinid fly of the genus *Blepharella* and a tiny wasp of the genus *Euplectrus*. *Blepharrella* females lay a great many very small eggs on young mango leaves, but they hatch only if swallowed by a caterpillar. The maggot feeds on the caterpillar's internal tissues, eventually killing it. The wasp larvae are gregarious, clinging to the outside of the caterpillar as they suck its fluids. Before fastening her eggs to the host's body, the female wasp prepares the way for her offspring by stinging the caterpillar to arrest its development. These two parasites so greatly decreased the shoot moth population that by 1988 the yield of mango fruits was 40 times greater than it had been before the mango shoot moth had been controlled.

The importation of prickly pear cactus into Australia late in the eighteenth century eventually resulted in an ecological calamity of awesome proportions. Grown as garden plants, the cacti escaped from cultivation and flourished. By 1900 they had overrun almost 16,000 square miles of rangeland, an area about twice the size of New Jersey. In 1925 they covered about 94,000 square miles, over 60 million acres, an area almost 12 times the size of New Jersey. It was obvious that their rapid spread would continue. The affected land was practically useless, about half of it so densely overgrown with tangles of the spiny cacti that it was literally impenetrable to people, cattle, and even kangaroos. In the Western Hemisphere, the only part of the world where cacti occur naturally, nothing even approaching such a plague of these plants had ever been seen. Held in check by their natural enemies, which did not, of course, occur in Australia, New World prickly pear cacti grow only in scattered clumps, never in huge, impenetrable tangles.

As recounted by J. K. Holloway,[12] the plague of cacti in Australia was eliminated by introducing several cactus-feeding insects from the New World. The most effective was a moth from South America that has no common name but has an appropriately descriptive generic name, *Cactoblastis*. These insects were introduced in 1925, and by 1937 they—but especially the *Cactoblastis* caterpillars—had become established and had destroyed the last dense stand of cactus. To this day, the cacti in Australia—their populations still controlled mainly by *Cactoblastis*—grow only in scattered clumps, and cattle, sheep, and kangaroos graze on once useless land.

Some entomologists would say that I am stretching the term "biological control" out of shape by applying it to the introduction of insects to dispose of dung. Bacteria and fungi are, of course, its ultimate decomposers, but insects also play an important role. Just how crucial they can be is shown by the introduction of beetles into Australia to rid the range of cattle droppings, as interestingly recounted by D. F. Waterhouse, an eminent Australian entomologist, in the April 1974 issue of *Scientific American*.[13]

In Europe and Africa, certain scarab beetles use the large, moist dung pats of buffalo and other cowlike grazers as food for themselves and their larvae. The scarab of ancient Egypt, for example, provides for each of its larvae by forming a large ball of cow or buffalo dung that it rolls to a suitable site and buries in the soil along with one of its eggs. Australian scarabs, however, are adapted to cope with only the small, dry dung pellets of kangaroos, the only large grazing animals that are native on that continent. When the English brought cows to Australia, there were no scarabs capable of helping to decompose their wet, sloppy dung. Cow droppings accumulated, drying out and persisting on the ground for months or even years before being decomposed by bacteria. This seriously impaired the productivity of pastures and rangeland by choking off the plants that cows eat. According to Waterhouse, the 30 million cows of Australia deposit about 360 million pats of dung per day. As you have probably guessed by now, this problem was alleviated by introducing into Australia from Africa and Europe several species of scarabs that were evolutionarily adapted to cope with the large, moist dung pats of cows.

"There's no such thing as a free lunch." This aphorism, a warning, is attributed to Milton Friedman, a Nobel laureate in economics. It is pertinent to almost all human endeavors, not excluding insect controls, almost all of which have hidden costs—side effects such as the extirpation of birds and the "creation" of new pests by insecticides. Even biological controls, often assumed to have no adverse effects, have had undesirable side effects of two sorts as astutely pointed out by Francis Howarth.[14] They can directly or indirectly affect one or more nontarget organisms and may, thereby, cause a "ripple effect" that may adversely affect an entire ecosystem. As we have seen, a parasitic fly introduced from Europe to control the gypsy moth has decimated populations of at least two of the giant American silk moths. But it is too soon to know if there will be a significant ecological ripple effect. In Hawaii the population of the voraciously predaceous mongoose increased after several scarab beetles introduced to remove cow dung from pastures became established and flourished. Prospering on this new and abundant insect food, the mongooses became so numerous that they threatened with extinction some of the few surviving native birds. In England the beautiful large blue butterfly (*Masculina arion*) was left homeless and became extinct when a purposely introduced virus decimated the population of rabbits, whose grazing had created the habitat favored by the butterfly.

The other side effect is that the introduced biological control organism may itself become a pest. As I write, people all over the United States, but especially in the northeastern states and southern Canada are being pestered by Asian ladybird beetles (*Harmonia axyridis*) that sneak into their homes, often by the thousands, to spend the winter—making do with buildings in lieu of crevices in rock formations and other places. Beginning in 1916, there were, according to E. S. Krasfur and his coauthors,[15] several attempts to establish this ladybird in the United States as a biological control agent. But it did not become established until 1988, after it had been introduced to control aphids on pecan trees.

The most amazing success of applied entomology is the use of the sterile male method not just to control, but to totally eradicate the screwworm fly, which literally devours the living flesh of animals, from all of North, Central, and South America. This was accomplished by releasing into the environment males that had been made reproductively sterile by exposing them to radioactivity. We will now explore how and why this innovative method was so successful.

The screwworm fly is a blow fly gone bad, a parasitic "outlaw" species that, in the maggot stage, feeds on the healthy flesh of living warm-blooded animals rather than on the decaying flesh of dead animals, as do many of its "lawful" relatives, notably the greenbottle and bluebottle flies. The glistening green or blue blow flies, about twice the size of a house fly, are conspicuous as they sponge up nectar from the blossoms of wild parsnip or wild carrot, Queen Anne's lace. The white, cone-shaped maggots are a less pleasing sight as they feed on the flesh of a dead animal. Nevertheless, repulsive as you may find them to be, these maggots are among the most important of the indispensable "undertakers" that recycle, return to the soil, the corpses of birds, mammals, reptiles, and other vertebrates, including humans.

The great eighteenth-century Swedish taxonomist Carolus Linnaeus said that three flies could consume the carcass of a horse as quickly as a lion, thereby paying dramatic although hyperbolic tribute to the ecological importance of blow flies as decomposers. At first glance, this seems impossible, but it is at least hypothetically possible if we factor in the prodigious reproductive potential of flies. Three female bluebottles lay about 900 eggs. If all their offspring were to live, which is admittedly very unlikely, they will after about 22 days be survived by over 20 million maggot grandchildren weighing about 500 pounds. If 50 percent of their food intake was devoted to growth, they would of necessity have eaten about a thousand pounds of carrion to attain their final weight, just about enough to dispose of all but the bones of the carcass of a horse. More realistically, about a quarter of their offspring might survive to become full-grown maggots, but, even so, they would have eaten about 250 pounds of carrion.

Like the screwworm, a few other flies of its family (Calliphoridae) have made the evolutionary transition from scavenging and carrion feeding to parasitism. Maggots of several species of *Protocalliphora* (they have no common name) live in the nests of both Old and New World birds and suck blood, like leeches, from the nestlings at night. In Africa south of the Sahara, the Congo floor maggot hides in dwellings and at night comes out to suck blood from people sleeping on the earthen floor. Maggots of the human bot fly, considered to be a calliphorid by Harold Oldroyd[1] but usually placed in a separate but closely related family, live just below the skin in the muscle tissues of humans and other animals. Their egg-laying behavior is remarkable. Instead of placing their eggs directly on the host animal, they capture another insect, usually a blood-sucking one, and attach a mass of their eggs to its body. When the unwitting porter lands on the skin of a warm-blooded animal, the eggs immediately hatch and the maggots burrow in.

These parasites seem comparatively harmless when compared to the New World screwworm or its Eurasian counterpart, known as the Old World screwworm. These gruesome parasites survive by feeding on the living flesh of a healthy, breathing animal. Carrion-feeding blow fly maggots often invade infected, puss-filled wounds, but unlike screwworms they feed on decaying rather than healthy flesh.

The medical profession has taken advantage of this latter behavior by using "lawful" blow flies to heal otherwise intractable infections. When a surgeon "debrides" a wound, he cannot help but cut away living as well as gangrenous, decaying flesh. But if he places specially raised, sterile maggots in the wound, they consume only the gangrenous tissue and leave almost every cell of the healthy tissue untouched. They also flood the wound with allantoin, an antibiotic substance. Maggot therapy, as it is called, was regularly used by physicians after World War I, was largely abandoned when modern antibiotics became available, but is now undergoing a renaissance as more and more bacteria become resistant to antibiotics.

The common name of the New World screwworm (which I will henceforth call simply the screwworm) refers to the encircling ridges on the body of the white maggot, which suggest the ridges, or threads, of a screw. The literal meaning of its scientific name, *Cochliomyia hominovorax*, is "the snail-like fly that devours people." Screwworms do infest people but much more often other warm-blooded animals: cattle, sheep, and other domestic animals, and wild animals ranging in size from deer to rats. Screwworm populations are generally low, only 100 to 200 flies per square mile, but that is enough to be seriously damaging. The egg-laying females are attracted to fresh wounds, which may be large or as small as a tick bite or a puncture from barbed wire. They are also attracted to the navel of new-born animals, to female genitalia—especially if injured in giving birth—and to nasal discharges. A female may lay 3,000 eggs in masses of from 10 to as many as 400 at the site of a wound. The eggs hatch in less than a day, and the newly hatched maggots cut their way into the healthy flesh, sever tiny blood vessels, and secrete a toxin that prevents healing. As wounds are enlarged by feeding screwworm maggots, they become ever more attractive to egg-laying females and are consequently enlarged ever more rapidly. The maggots feed for about eight days, grow to be about 2/3 inch long, and then drop to the ground, burrow into the soil, form a puparium, and emerge as adults after about a week.

In 1934, according to W. E. Dove[2] of the US Bureau of Entomology and Plant Quarantine, about 1.3 million cattle were infested with screwworms in the southeastern states, and about 12 percent of them ultimately died. In 1935 there were 55 confirmed infestations in humans and enough unconfirmed infestations to raise the probable total to more than 100. Dove's accounts of the infestations in humans are gruesome and horrifying:

> J.B., age 3 years 6 months, Sanford, Florida, injured skin about size of dime resulting from fall off bicycle was treated at home, but failed to heal. Three or four days later a marked swelling, some discoloration, a hemorrhage and an odor were present. Dr. B. A. Burkes, of Winter Park, lanced the swelling, which was found to contain 10 or 12 larvae.

> J.M.C. of Franklin, Texas, a farmer with chronic sinusitis, slept on a porch about noon on July 15, 1935. The following day he had a tickling sensation in his nose, which was followed by pain in the left sinus. He thought the pain was caused by a decayed tooth and had it pulled. The pain became more intense and . . . during the following nine days a total of 385 larvae dropped from both his nostrils. J.M.C. was successfully treated by saturating cotton with chloroform and packing it into the nasal cavities for short periods. He suffered a great deal of pain, had hemorrhages from the nose, and was delirious several times during the course of the infestation.

> N. McK. . . . age about 70 years, lived alone about 14 miles north of Branford, Texas. In June 1935 she fell and was unable to move. When discovered several days later and placed in a bed, Dr. P. C. Farnell found that screwworms had entered the vagina and were active in the sacral regions and the navel. She was treated with chloroform and olive oil. It was estimated that about 1 quart of larvae were obtained from the wound. Treatment was given on June 14 and the woman died June 15, 1935.

Human infestations, although hideous and potentially dangerous, were rare, and could be avoided by taking sensible precautions. But cattle and other animals could not take precautions and were virtually helpless if attacked by screwworms. The monetary loss to cattle growers was great—in Florida alone, $20 million in 1959, about $160 million in current dollars. Some of the loss could be avoided by timing the branding, dehorning, castration, and other surgical management procedures so as not to wound cattle during the flies' season of abundance. But a great many wounds attractive to egg-laying screwworms were inevitable, such as tick and fly bites and scratches from thorns or barbed wire. The only available method of control was the labor-intensive and expensive procedure of examining every cow and applying insecticidal ointment to maggot infestations and to as yet uninfested wounds.

In this case, trying to eliminate a very small population of adult flies spread over a vast area with a modern synthetic insecticide—DDT, methoxychlor, malathion, or any of the others—was obviously impracticable. But Edward Knipling,[3] chief of the Entomology Research Branch of the US Agricultural Research Service, had the imagination and foresight to go beyond the conventional wisdom of the time: the simplistic view that the best way to control insects is to kill them with an insecticide. He and his colleagues went back to the basics, looking for a weak spot in the screwworm's life cycle by learning as much as they could about its way of life. One of their first research projects was a study of the sex life of the screwworm fly. This project probably sounded laughable to the uninitiated, but, fortunately, this was before William Proxmire was elected to the US Senate and began to present his famous Golden Fleece Award, meant to ridicule research that he, trained only in business administration, considered to be a waste of the taxpayers' money. It was a good thing that he was not there to interfere, for the knowledge gained from this study ultimately resulted in the total extermination of screwworms from the United States south to Panama, thereby saving and continuing to save hundreds of millions of dollars every year.

The early results of this research revealed the interesting fact that, unlike many other flies, female screwworms mate only once in a lifetime. With her spermatheca filled with enough sperm to fertilize all the eggs she can lay, she has no further need of a male and devotes the rest of her life to locating host animals on which to lay her eggs. The males, on the other hand, are promiscuous, driven by their genes to enhance their evolutionary fitness by fathering as many progeny as possible.

This simple scenario stimulated Knipling's imagination, prompting him to ask some questions that no one else had ever asked: Could male screwworms be artificially sterilized, and would they afterward retain their promiscuous mating behavior? Would mating with a sterile male turn off a female's interest in sex, thereby condemning her to lay only unfertilized eggs? And finally, could a screwworm population be reduced or even exterminated by releasing into their environment sterile males that would compete for females with the wild, fertile males? The answers to all three of these questions turned out to be a resounding yes that stunned economic entomologists all over the world.

In 1955 A. H. Baumhover and several coauthors[4] recounted the extensive research that conclusively demonstrated that Knipling's idea was sound—that a population of screwworms could be eradicated by what has come to be known as the sterile male technique. The first step was to devise a way of rearing large numbers of screwworms in the laboratory. The flies grew well on a diet concocted from horse meat and blood, and eventually screwworms were raised by the hundreds of millions in "screwworm factories." Sterilization of both males and females was accomplished by exposing the puparia to gamma rays emitted by radioactive cobalt-60. The first test of the sterile male method, done with flies in a large cage in the laboratory, boded well for the success of the method under natural conditions. As Knipling[5] reported, "when sterile and fertile males in a ratio of 4 to 1 were introduced in a cage of virgin females about 80 percent of the females deposited eggs that did not hatch."

The first field trial was on Sanibel Island, about two miles off the west coast of Florida. Sterile males were released at the rate of 100 per square mile per week. The fertility of egg masses, collected from penned goats, declined rapidly, and after eight weeks of releases, the screwworm population on the island appeared to have been eradicated. But on the twelfth week a single fertile egg mass was collected. It had almost certainly been deposited by a fertile female from the flourishing mainland population that had flown or been blown the short distance to the island. Obviously, the efficacy of the method could be demonstrated with absolute certainty only on an island well removed from other sources of infestation.

The 170-square-mile island of Curaçao, in the Netherlands Antilles 40 miles north of the coast of Venezuela, was home to about 25,000 goats and some 5,000 sheep, which roamed through the brush and cactus largely unattended and were heavily infested with screwworm maggots. This was an ideal place to test the sterile male method.

Before, during, and after the program of releasing sterile males on Curaçao, the screwworm population of the island was monitored by checking goats in 11 fenced enclosures, strategically placed along the whole length of the island, for the presence of egg masses. "Each week," wrote Baumhover and his colleagues,[6] "two goats in each pen were [superficially] wounded and artificially infested with approximately 100 newly hatched screw-worm larvae. When the larvae were 3 days old, the wounds were fumigated with benzol to destroy the larvae." The wounds, almost irresistibly attractive to egg-laying flies, were a sensitive "lightning rod" that kept track of the screwworm population.

Before any sterilized males were released, 288 egg masses, each consisting of many eggs that were all fertile, were found on penned goats during a period of 13 days. Beginning on August 9, 1954, an average of 400 sterile males and 400 sterile females per square mile were released from an airplane over the entire island approximately once a week. The number of fertile egg masses quickly declined, and after the sixth release, in the middle of September, the last three fertile egg masses were found. Sterile male releases continued for about the next 16 weeks, until January 6, 1955, and during that time not one fertile egg mass and only two sterile egg masses were found. The screwworm had been eradicated from Curaçao and has not returned.

At this point, the big question was whether or not screwworms could be eliminated from the US mainland. There were two separate populations, one in the Southwest and another in the Southeast. Screwworms first appeared in the Southeast in 1933, probably brought in on infested livestock from the Southwest. In cold winters the southeastern population survived only on the southern two-thirds of the Florida peninsula, but in spring and summer they moved northward, often transported on infested cattle, and occupied most of the Southeast and sometimes appeared as far north as Illinois. Raymond Bushland and Edward Knipling—who in 1992 shared the $200,000 World Food Prize for their development of the sterile male technique—realized that in winter the southern part of the Florida peninsula is, for screwworms and many other insects, an island surrounded by the sea on three sides and by cold weather on the north. During the winter of 1958–59, the screwworm was eradicated from peninsular Florida, thereby freeing the entire Southeast of this pest. Hundreds of millions of sterile males and females were dropped from aircraft on 85,000 square miles of southern Florida. (They bothered no one because they were released at the rate of only 200 per square mile per week.) The last screwworm ever to be seen in the Southeast was found in Florida on February 19, 1959. This sterile male release program cost a seemingly prohibitive $10 million, but at the time the yearly loss to screwworms was $30 million just in Florida.

The eventual eradication of these pests from the Southwest freed the whole country of them and has over the years saved our economy billions of dollars. In the Southwest, sterile males had to be dropped on a wide barrier zone, a "no-man's-land," every year to keep flies from Mexico from moving back into the area that had been freed from them. At first the no-man's-land, extending along the Mexican border, was about 2,000 miles long. Although sometimes delayed by political problems, the screwworm has now been exterminated south to the 60-mile-wide Isthmus of Panama, where a narrow and therefore inexpensively maintained barrier zone is a "cork in the neck of the bottle" that holds the screwworms of South America at bay.

The sterile male technique is not a panacea. At best it will be useful for the control of only a limited number of insects. It is, however, a very attractive procedure because it alone offers the hope of totally eradicating a pest from a large area, something that has yet to be accomplished with any insecticide. Sterile males have been useful in several instances. For example, they were an important part of a program to eradicate a recently introduced infestation of the Mediterranean fruit fly in California; to control—but not exterminate—codling moths in two apple orchards in British Columbia, Canada; and to eliminate locally a mosquito from a rice-growing village in Burma.

Another way to prevent female insects from laying fertile eggs is to annihilate the males of their species. This male annihilation technique was used to exterminate a destructive pest from a small tropical island in the western Pacific. The inhabitants of Rota, 37 miles north of Guam in the Marianas archipelago, were plagued by a very large and destructive population of oriental fruit flies (*Bactrocera dorsalis*). The maggots, which grow rapidly—attaining full size in only about a week—must, of course, eat ravenously. They tunnel within and feed on the fruits of at least 173 species of plants, turning the flesh of the fruit into a mushy mess. On Rota they infested breadfruits, mangoes, papayas, citrus, and some other fruits. This was a serious problem, and no simple remedy was at hand. But the people of Rota were in luck. A team of US Department of Agriculture entomologists led by L. F. Steiner[7] chose Rota as a testing ground for a then novel way of controlling insects, by attracting males to an insecticidal booby trap by means of an attractant. It does not matter that the females are not killed because without males to inseminate them they can lay no fertile eggs and the population rapidly dwindles. The test on Rota was a success; the method worked. It killed males by the hundreds of thousands and once and for all eradicated the oriental fruit fly from the entire island.

Choosing an effective attractant was easy. Methyl eugenol, for reasons still not understood, is irresistibly attractive to several species of fruit flies, including the oriental fruit fly. Early in the twentieth century, F. M. Howlett[8] made a serendipitous discovery. While he was experimenting in Pusa, India, trying to find a way to control the destructive peach fruit fly (*Dacus zonatus*), he found that methyl eugenol is a potent attractant for this insect. Howlett hoped to alleviate the injury done to peaches and mangoes by inducing females to lay their eggs somewhere other than on these fruits. His idea was to attract them by imitating the smell of ripe peaches or mangoes. After failed attempts with many substances, he heard that a "neighbor had been troubled by some kind of fly settling on him . . . when he was using oil of citronella, sprinkled on his handkerchief, as a mosquito deterrent." As soon as Howlett exposed a handkerchief wetted with oil of citronella near a peach orchard:

> . . . it became evident that the smell exercised an extraordinarily powerful attraction. In less than half an hour the handkerchief, lying in a crumpled heap, was almost hidden by a crowd of *D. zonatus*, and presented a very striking appearance. I jumped at once to the conclusion that the economic problem of how to destroy female fruit-flies had found an easy solution, but on examination it was soon apparent that all the flies on the handkerchief were males; they almost refused to leave the neighbourhood of the handkerchief, and a considerable number of them followed me home when I removed it.

Howlett later isolated and identified methyl eugenol as the attractive component of the chemically complex oil of citronella. This chemical compound has such an overpowering effect on male oriental fruit flies, reported Roy Cunningham, that they will drink it in its pure form "until they fill their crops and die." Forty-three species of fruit flies are now known to be methyl eugenol addicts, and more are likely to be added to the list.

The reason why methyl eugenol is so attractive to fruit flies is still the subject of conjecture. Some have argued that male fruit flies are attracted because this substance has a fortuitous chemical resemblance to the females' sex attractant pheromone, but an observation that belies this hypothesis is that it also attracts females under certain circumstances. Others suggested that it resembles the odor of a fruit in which these flies breed. Methyl eugenol has, indeed, been found in 25 plants of several different families, but none of them are breeding hosts for fruit fly larvae.

During the eradication campaign on Rota, methyl eugenol lured male oriental fruit flies to booby traps, lethal baits. These were small fiberboard squares that had been soaked in methyl eugenol laced with a tiny amount of naled, an insecticide that kills insects if it is absorbed through the skin and also if it is ingested. An airplane dropped poison baits on the island 15 times at the rate of 125 per square mile at intervals of about two weeks over a period of about eight months. A miniscule amount of naled was released into the environment, a little more than a tenth of an ounce per acre in each drop. Naled had some residual effects; baits attracted and killed flies for two months. But after that time the insecticide had been biodegraded and made harmless.

The population of flies was monitored by catching males in traps baited with methyl eugenol and by collecting fruits and noting the emergence of adults from them. The results of the program were spectacular. During the two-week period before the first drop, the traps averaged a catch of 262 flies per day. After the first drop the average catch had already been reduced by 93 percent, about 18 flies per trap per day. And the catch steeply declined with each succeeding drop until the sixth one, after which no more flies were caught. The last adults to emerge from a fruit came out of mangoes six months after the first drop. The oriental fruit fly had been eradicated from Rota. Since then oriental fruit flies have been eradicated from other islands and incipient infestations, resulting from the unintentional importation of this fly, have been eliminated in California.

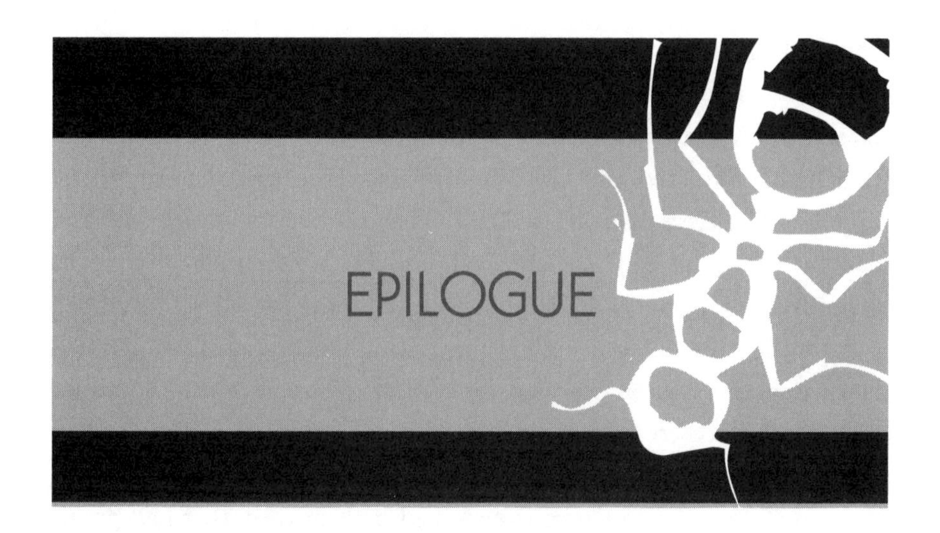

EPILOGUE

Evolution, driven by natural selection, has created over 300,000 different kinds of plants and at least 1,200,000 different kinds of animals, 900,000 of which are insects. As a group, these insects are essential to our survival. In virtually all land and freshwater ecosystems, they perform indispensable services, ranging from pollinating plants to recycling dead plants and animals. Without them, these ecosystems—including agricultural ecosystems—would collapse, and life as we know it could not persist. Less than 2 percent of these insects—doing what comes naturally—eat our crops, transmit diseases, and come into conflict with us in other ways. Only these few can rationally be considered pests. Nevertheless they, including the few we have met, were molded by natural selection, as were all other organisms that over time became adapted to cope with the same basic problems faced by all insects. From these pests we can learn a great deal that applies to all insects and, to a lesser extent, to all life.

We have seen, for instance, indisputable examples of evolution in action. Within months of their first contact with DDT, natural selection had eliminated the vast majority of house flies, except a tiny minority that were resistant to DDT. They survived, and their descendants, all resistant to DDT, are the house flies we see today. By means of natural selection, two different species of corn rootworms developed two totally different ways of circumventing crop rotation, which had had these two beetles stymied for almost a hundred years. Natural selection favored an infinitesimal minority of peppered moths that had a mutant gene or genes that changed their color, altering their camouflage to conform with the soot-darkened bark of trees that had previously been light in color. In areas polluted by smoke, the dark form of this moth soon replaced the light form. Finally, there is convincing evidence that a fruit fly, the apple maggot, is splitting into two species: one that feeds on the fruits of its native host, hawthorn, and another that has adapted to a new host, the apple, which was introduced into North America just a few hundred years ago.

Now that we know how insects, and indeed all animals, originated, we are faced with the question of in what ways they have become adapted to survive and to get food and other resources from their environment. In a delightful little book on ecology, Paul Colinvaux[1] writes that the animals now on Earth are the survivors of an evolutionary struggle in which they "slowly changed over the years so that the ones we have are beautifully fitted to the lives they must lead. They have evolved to find food, to survive hazards, and to make babies under the local circumstance or 'environment.'" In short, any animal's life is ruled by three imperatives: it must eat and grow, avoid being eaten, and produce offspring.

About 90 percent of the plant-feeding insects are adapted to feed on just a few plants of one family or of two or three related families.[2] Among them are corn rootworms, apple maggots, codling moths, and Hessian flies, all of which would starve to death rather than feed on a plant that is not on their "menu." The cabbage worm and diamond-back moth—and probably all "host-specific" insects—recognize their proper host plants by the taste or odor of their nonnutrient "secondary" chemicals. Most of these chemicals—thousands are scattered throughout the plant kingdom—are toxic and have passed the muster of natural selection because they protect plants against insects. Some insects evolved ways of detoxifying the "chemical warfare agents" of a particular group of plants and then came to rely on the flavor of their secondary compounds to recognize them as being safe to eat. Host-specific insects are wedded to their food plant by a constellation of adaptations. Not only can they cope with the plant's defenses; they are also synchronized with its seasonal cycle; their digestive systems are fine-tuned to best use it as a food; and their camouflage conforms with the plant's appearance.

Insects, like human infants, rats, birds, spiders, and probably all other animals, can select a balanced diet from an array of separate nutrients. My colleagues and I[3] demonstrated that the plant-feeding corn earworm, a seed-eating grain beetle, and an omnivorous cockroach have this ability. Corn earworms, as you have seen, select an optimal balance of sugar and protein in the laboratory from artificial diets. In nature this ability is an important dimension of insect feeding behavior. Corn earworms and grain beetles ingest optimal proportions of the vitamin-rich germ and the starchy part of corn kernels and wheat seeds. If given a mix of several plant species, grasshoppers eat them in an optimal mix and grow better than if fed only one of them. It is a fair assumption that other insects behave similarly. Female mosquitoes, for example, may ingest an optimal proportion of blood and nectar, bees an optimal proportion of nectar and pollen, and predaceous insects may improve their diet by eating a balanced mix of several different species of prey.

Surviving winter is a problem for all insects of the Arctic and temperate zones. The great majority survive in the hibernation-like state called diapause. They stop developing, become resistant to freezing, and lower their metabolic rate enough to survive the long winter on the fat they have stored in their bodies. Diapause may occur in any developmental stage. The subject of our consideration of diapause, the bagworm, diapauses as an egg, as do aphids and the gypsy moth. Codling moths and Japanese beetles diapause in the larval stage, cabbage worms and corn earworms as pupae, and chinch bugs and some mosquitoes in the adult stage. Diapause may end and development resume at any time in spring or even summer, but all species terminate diapause more or less in synchrony and at a time favorable to themselves: at the most propitious time for mating; when the temperature is likely to be favorable; when the flowers from which they sip nectar are in blossom; or, in the case of parasites, when their hosts are available.

All insects are in danger of being eaten by predators such as bats, mice, birds, spiders, or other insects. But they have many ways of protecting themselves. Night-flying moths have ears and take evasive action when they hear the echolocation cries of bats. Honeybees protect their colony by stinging intruders. Black swallowtail butterflies have several defenses. Small larvae are ignored by birds because they look like bird droppings. Pupae are likely to go unnoticed because they are camouflaged to blend in with either the green of summer or the browns and grays of winter. Adults are edible but are passed up by birds because they mimic a toxic butterfly.

But no defense is perfect, and few insects survive the onslaught of predators and parasites. And that is a very good thing. Think about this. If only 2 survive of a gypsy moth's 400 eggs, a house fly's 500, or a corn earworm's 1,000 to become reproducing adults, they will have replaced their parents, and the population will remain stable from year to year. But if only two extra survive, a total of four, the population will double with each generation, and there will eventually be a horrendous population explosion. After only 10 house fly generations—a year—the house fly population will have multiplied by an ecologically and economically disastrous factor of 500 times! Clearly, 498 of a pair's

500 offspring must die before they can reproduce. Fortunately for the well-being of Earth's ecosystems, population explosions are rare, in large part because predators and parasites are doing their job. What happened when insecticides killed predators and parasites in apple orchards is convincing evidence of their indispensable ecological role. Without predators and parasites to control them, small and economically insignificant populations of some apple-feeding insects that were immune to the insecticides soared out of control and became an economic threat.

Reproduction is the ultimate evolutionary imperative because it passes genes from generation to generation, the genes that encode "information" that has passed the muster of natural selection and is a "blueprint" for survival. An amazing multitude of different anatomical, physiological, and behavioral adaptations facilitate reproduction. Not least among them are the many different ways in which males and females come together. For example, many species attract mates with sex pheromones, as do bagworms and gypsy moths. Males of some species lie in wait for females where they lay their eggs, as do apple maggots, or where they come to feed, as do tsetse flies. Male and female mosquitoes, like many other flies, meet in large swarms that form above distinctive and conspicuous landmarks.

Evolution has programmed all animals to reproduce so as to maximize the number of their offspring that will survive to become reproducing adults. Aphids and tsetse flies exemplify two very different strategies for achieving this end. Aphids play the numbers game, producing many progeny and gambling that a few will, by luck, survive. Tsetse flies produce very few progeny but increase their survival rate by giving them extensive care.

Insects that produce many progeny—aphids, corn earworms, bagworms, Hessian flies—invest relatively little time and energy in them. A female mayfly may dump all of her eggs in the water in one big mass. But in many of these species there is a modicum of parental care.

Before abandoning them, host-specific plant-feeders lay their eggs on plants suitable for their offspring. Female bean leaf beetles bury their eggs in the soil at the base of a bean plant; apple maggots insert them one at a time into apples; and black swallowtails glue them one by one to the leaves of parsley or a related plant. Solitary bees give more care and, consequently, can lay fewer eggs. In a burrow or nest they lay a single egg on a mass of pollen and nectar sufficient to feed the larva until it is ready to pupate. Solitary wasps do much the same but stock their nests with insects or spiders.

The ultimate caregivers are the tsetse fly, the sheep ked, and several small, wingless flies that live in the fur of bats and suck their blood. Their "strategy" is not to lay eggs but, rather, to bear a few larvae, perhaps only five or six in a lifetime, retain them in their bodies during a period of gestation, and give birth to them when they are full grown and ready to molt to the pupal stage. A few other insects lay eggs, but feed and care for their growing larvae until they are ready to pupate. When a pair of burying beetles finds a dead mouse or some other small animal, they bury it, the female lays a few eggs in the soil nearby, and both male and female care for the larvae until they are ready to pupate, feeding them by regurgitating predigested carrion. Edward O. Wilson[4] writes that burying beetles interact with their larvae much as birds interact with their nestlings. Among the more than 4,000 species of cockroaches, the great majority of which are not household pests, we see an evolutionary progression toward increasing parental care. Many species enclose their eggs in a protective case that they carry at the end of the abdomen until the eggs are ready to hatch. Some go further and retain their nymphs in a brood chamber, much as a kangaroo shelters her joey in her pouch. As quoted by Louis Roth and Edwin Willis, G. N. Wolcott said of another cockroach, ". . . the mother broods over her young, and together they sally forth at night in search of food, until they are of such a size as to mingle with their elders."[5]

The 900,000 known insect species—several million are not yet known to science—are, we have noted, indispensable members of virtually all of the land and freshwater ecosystems on Earth. Without them, life on Earth as we know it and probably humanity itself could not survive, since they have the many indispensable ecological responsibilities already mentioned. That said, we must face the fact that a small number of insects—less than 2 percent of them—are pests that eat stored products such as grain, damage our crops, and transmit diseases. According to Robert A. Metcalf and Robert L. Metcalf,[6] insects cost the American economy over 14 billion dollars in 1988—that would be much more in today's inflated dollars.

What can we do to decrease this loss or—probably more realistically—prevent its increase? Since its beginnings in the nineteenth century, applied entomology has learned much about what to do—and at least equally important—what not to do. The cure-all, or panacea, approach is to find out which insecticides kill the insects and then spray. The other approach is the modern concept of Integrated Pest Management: know your enemy, discover its weaknesses, and in the light of that knowledge, do what will work best, perhaps a noninsecticidal control alone or in combination with the judicious use of an insecticide.

Insecticides are sometimes the best way to control insects. For example, the worldwide campaign to eradicate malaria almost succeeded, but, when the mosquitoes became resistant to DDT, it failed and was followed by a massive resurgence of malaria. Nevertheless, DDT had kept malaria at bay for 30 years, preventing much suffering and many deaths. But, like medications, insecticides can have unwelcome side effects, usually when they are misused or overused. We have seen that applying DDT and other insecticides to apple orchards "created" new pest insects and mites by killing off their predators and parasites. The misuse of dieldrin in a futile attempt to prevent Japanese beetles from entering Illinois from Indiana had horrendous environmental effects. Indiscriminately spraying this very toxic substance on fields, farmsteads, and towns killed pets, livestock, and almost eradicated squirrels and so many songbirds that Rachel Carson titled her exposé of the misuse of insecticides *Silent Spring*.[7]

We have explored several of the many ways of controlling or even eradicating pest insects without insecticides or with their judicious use. Tsetse flies have been eliminated as a problem by inducing them to land on "fake" cows laced with insecticide. The French wine industry was rescued from the grape phylloxera by grafting the susceptible European wine grapes on the rootstocks of resistant American grapes. Japanese beetle larvae can be controlled by spreading the spores of the milky disease—harmless to people and animals not related to the Japanese beetle—on lawns in areas that are infested by the beetle. After the substitution of soybeans for wheat in the midwestern cropping system, chinch bugs were no longer a problem. Hessian flies are controlled by planting wheat after the egg-laying females have died. And finally, the screwworm fly has been totally eradicated from the United States south to Panama by releasing sterile males to compete with fertile males.

We know much more about pest insects—those that conflict with our interests—than about the great majority of other insects. Because they affect us economically, we spend more to support research on them than on insects we consider to be harmless. That is as it should be—at least from an economic point of view. Fortunately, what we learn about pest insects applies to all insects, and teaches us a great deal about the role of insects in the worldwide web of life, upon which we depend for our very existence.

INTRODUCTION

1. Larry P. Pedigo, *Entomology and Pest Management* (New York: Macmillan, 1989), p. 30.

2. Stephen A. Forbes, *The Insect, the Farmer, the Teacher, the Citizen, and the State* (Urbana: Illinois State Laboratory of Natural History, 1915), p. 2.

CHAPTER 1. THE MOST DANGEROUS INSECTS: MOSQUITOES

1. Andrew Spielman and M. D'Antonio, *Mosquito: A Natural History of Our Most Persistent and Deadly Foe* (New York: Hyperion, 2001), p. xx.

2. C. O. Farquharson, "*Harpagomyia* and other Diptera fed by *Crematogaster* ants in S. Nigeria," *Transactions of the Entomological Society of London* (1918): xxix–xxxix.

3. Frederick Knab, "The swarming of *Culex pipiens*," *Psyche* 13 (1906): 123–33.

4. William R. Horsfall, *Mosquitoes; Their Bionomics and Relation to Disease* (New York: Ronald Press, 1955), p. 522.

5. Patrick Manson, "On the development of *Filaria sanguinis hominis*, and on the mosquito considered as a nurse," *Journal of the Linnaean Society of London, Zoology* 14 (1878): 304–11.

6. Spielman and D'Antonio, *Mosquito,* p. 195.

7. Robert J. Novak, conversation with author, 2003.

8. Ibid.

9. Spielman and D'Antonio, *Mosquito,* p. 200.

10. *Malaria Control on Impounded Water* (Washington, DC: US Public Health Service and Tennessee Valley Authority, 1947), p. 3.

11. William R. Horsfall, *Medical Entomology* (New York: Ronald Press, 1962), p. 339.

12. Spielman and D'Antonio, *Mosquito,* p. 129.

13. Ibid., pp. 157–78.

14. Robert L. Metcalf, "Insecticides in Pest Management," in *Introduction to Pest Management,* ed. R. L. Metcalf and W. H. Luckmann (New York: John Wiley and Sons, 1994), pp. 245–314.

15. Ibid., p. 252.

CHAPTER 2. EVOLUTION IN ACTION: HOUSE FLY

1. George S. Graham-Smith, *Flies in Relation to Disease: Non-Bloodsucking Flies* (Cambridge: Cambridge University Press, 1914), p. 19.

2. Luther S. West, *The Housefly* (Ithaca, NY: Comstock Publishing Group, 1951), p. 275.

3. Thomas R. Dunlap, *DDT* (Princeton, NJ: Princeton University Press, 1981), p. 3.

4. Ibid.

5. Anthony W. A. Brown and R. Pal, *Insecticide Resistance in Arthropods* (Geneva: World Health Organization, 1971), pp. 300–301.

6. James G. Sternburg, Clyde W. Kearns, and Herbert Moorfield, "DDT-dehydrochlorinase, an enzyme found in DDT-resistant flies," *Journal of Agricultural and Food Chemistry* 2 (1954): 1125–30.

7. Mark L. Winston, *Nature Wars: People vs. Pests* (Cambridge, MA: Harvard University Press, 1997), p. 14.

8. R. R. Parker, "Dispersion of *Musca domestica* Linnaeus under city conditions in Montana," *Journal of Economic Entomology* 9 (1916): 325–54.

9. F. C. Bishopp and E. W. Laake, "The dispersion of flies by flight," *Journal of Economic Entomology* 12 (1919): 210–11.

10. Robert L. Metcalf and Robert A. Metcalf, *Destructive and Useful Insects,* 5th ed. (New York: McGraw-Hill, 1993), p. 21.44.

11. Thomas H. Frazzetta, *Complex Adaptations in Evolving Populations* (Sunderland, MA: Sinauer Associates, 1975), p. 2.

12. Vincent B. Wigglesworth, *The Principles of Insect Physiology* (Cambridge: Cambridge University Press, 1972), pp. 268–69.

13. Charles Darwin, *On the Origin of Species by Means of Natural Selection or the Preservation of Favoured Races in the Struggle for Life* (London: John Murray, 1859). Refer to entire book.

CHAPTER 3. WHAT DARWIN WISHED HE KNEW: *DROSOPHILA*

1. Donald J. Borror, Dwight M. DeLong, and Charles A. Triplehorn, *An Introduction to the Study of Insects*, 5th ed. (Philadelphia: Saunders, 1981), pp. 609–10.

2. Douglas J. Futuyma, *Evolutionary Biology*, 2nd ed. (Sunderland, MA: Sinauer, 1986), pp. 241–42, 303–305.

3. Michael Ashburner, "Entomophagous and other bizarre Drosophilidae," in Michael Ashburner, H. L. Carson, and J. N. Thompson Jr., *The Genetics and Biology of Drosophila*, vol. 3a (London: Academic Press, 1981). Consult all of chap. 10, pp. 398–421.

4. Warren P. Spencer, "Collection and Laboratory Culture," in *Biology of Drosophila*, ed. M. Demerec (New York: Hafner, 1950), p. 49.

5. M. Ashburner and J. N. Thompson Jr., "The Laboratory Culture of *Drosophila*," in *The Genetics and Biology of Drosophila,* vol. 2a, ed. M. Ashburner and T. R. F. Wright (London: Academic Press, 1978), pp. 2–29.

6. Code of Federal Regulations, www.fda.gov/ora/compliance_ref/cpg/cpgfod/cpg585-890.htm (January 16, 2004).

7. C. Darwin, *Variations of Animals and Plants under Domestication,* reprinted in *The Works of Charles Darwin*, ed. P. H. Barrett and R. B. Freeman (New York: New York University Press, 1988). See note 9.

8. Gregor Mendel, "Versuche über Pflanzenhybriden" (Research on plant hybrids), *Verhandlungen der naturforschungs Verein Brünn* 4 (1866): 3–47.

9. Ronald W. Clark, *The Survival of Charles Darwin* (New York: Random House, 1984), pp. 169–74.

10. Monroe W. Strickberger, *Evolution*, 2nd ed. (Sudbury, MA: Jones and Bartlett, 1996), p. 483.

11. Ibid., p. 173.

12. Ibid., p. 35.

13. Clark, *The Survival of Charles Darwin*, pp. 215–16.

14. Ibid., pp. 216–17.

15. Ibid., pp. 241–42.

16. Warren P. Spencer, "Collection and Laboratory Culture," in Demerec, *Biology of Drosophila*, pp. 541–47.

17. Herman J. Muller, "Thomas Hunt Morgan, 1866–1945," *Science* 103 (1946): 550–51.

18. Clark, *The Survival of Charles Darwin*, p. 257.

19. Ibid., p. 261.

20. Muller, "Thomas Hunt Morgan," p. 551.

CHAPTER 4. NATURAL SELECTION OUTFLANKS FARMERS: CORN ROOTWORMS

1. Paul C. Mangelsdorf, *Corn: Its Origin, Evolution, and Improvement* (Cambridge, MA: Harvard University Press, 1974), p. 1.

2. F. M. Webster, "On the probable origin, development, and diffusion of North American species of the genus *Diabrotica*," *Journal of the New York Entomological Society* 3 (1895): 158–66.

3. Terry F. Branson and James Krysan, "Feeding and oviposition behavior and life cycle strategies of *Diabrotica*: an evolutionary view with implications for pest management," *Environmental Entomology* 10 (1981): 826–31.

4. Rosanna Giordano, Jan Jackson, and Hugh Robertson, "The Role of *Wolbachia* bacteria in reproductive incompatibilities and hybrid zones of *Diabrotica* beetles and *Gryllus* crickets," *Proceedings of the National Academy of Sciences* USA 94 (1997): 11439–44.

5. Michael E. Gray and William H. Luckmann, "Integrating the cropping system for corn insect pest management," in *Introduction to Pest Management*, ed. Robert L. Metcalf and William H. Luckmann (New York: John Wiley, 1994), p. 516.

6. H. C. Chiang, "Survival of northern corn rootworm eggs through one and two winters," *Journal of Economic Entomology* 58 (1965): 470–72.

7. James L. Krysan, Jan Jackson, and A. C. Lew, "Field determination of egg diapause in *Diabrotica* with new evidence of extended diapause in *D. barberi* (Coleoptera: Chrysomelidae)," *Environmental Entomology* 13 (1984): 1237–40.

8. Eli Levine, H. Oloumi-Sadeghi, and J. R. Fisher, "Discovery of multiyear diapause in Illinois and South Dakota northern corn rootworm (Coleoptera: Chrysomelidae) eggs and incidence of the prolonged diapause trait in Illinois," *Journal of Economic Entomology* 95 (1992): 262–67.

9. Eli Levine, J. L. Spencer, S. A. Isard, D. W. Onstad, et al., "Adaptation of the western corn rootworm to crop rotation: evolution of a new strain in response to a management practice," *American Entomologist* 48 (2002): 94–107.

CHAPTER 5. HOW A SPECIES BECOMES TWO SPECIES: FRUIT FLIES

1. A. L. Quaintance and E. H. Siegler, *The More Important Apple Insects* (Washington, DC: US Department of Agriculture Farmers' Bulletin 1270, 1922), pp. 14–15.

2. James R. Carey, "The Mediterranean fruit fly invasion of Southern California," in *The Medfly in California: Defining Critical Research,* ed. J. G. Morse, R. L. Metcalf, J. R. Carey, and R. V. Dowell (Riverside: College of Natural and Agricultural Sciences, University of California, 1994), pp. 71–91.

3. E. F. Boller and Ronald J. Prokopy, "Bionomics and management of *Rhagoletis,*" *Annual Review of Entomology* 21 (1976): 223–46.

4. Ronald J. Prokopy and Jorge Hendrichs, "Mating behavior of *Ceratitis capitata* on a field-caged host tree," *Annals of the Entomological Society of America* 72 (1979): 642–48.

5. Chris T. Maier and G. P. Waldbauer, "Dual mate-seeking strategies in male syrphid flies. (Diptera: Syrphidae)," *Annals of the Entomological Society of America* 72 (1979): 54–61.

6. Ronald J. Prokopy and Daniel R. Papaj, "Behavior of flies of the genera *Rhagoletis*, *Zonosemata*, and *Carpomya* (Trypetinae: Carpomyina)," in *Fruit Flies (Tephritidae), Phylogeny and Evolution of Behavior,* ed. M. Aluja and A. L. Norrbom (Boca Raton, FL: Chemical Rubber Company, 2000), pp. 219–52.

7. Ronald J. Prokopy, "Evidence for a marking pheromone deterring repeated oviposition in apple maggot flies," *Environmental Entomology* 1 (1972): 326–32.

8. Ibid.

9. F. Diaz-Fleischer, D. R. Papaj, R. J. Prokopy, et al., "Evolution of fruit fly oviposition behavior," in Aluja and Norrbom, *Fruit Flies,* pp. 812–41.

CHAPTER 6. GUARANTEEING DESCENDANTS: THE NUMBERS GAME—APHIDS

1. Robert E. Snodgrass, *Insects, Their Ways and Means of Living* (New York: Smithsonian Institution Series, 1930), p. 278.

2. Nancy A. Moran, "The evolution of aphid life cycles," *Annual Review of Entomology* 37 (1992): 321–48.

3. L. J. Pickett, L. J. Wadhams, C. M. Woodcock, et al., "The chemical ecology of aphids," *Annual Review of Entomology* 37 (1992): 67–90.

4. A. F. G. Dixon, "Evolution and adaptive significance of cyclical parthenogenesis in aphids," in *Aphids, Their Biology, Natural Enemies, and Control*, vol. A, ed. A. K. Mink and P. Harrewijn (Amsterdam: Elsevier, 1987), p. 294.

5. Ursula Goodenough, *The Sacred Depths of Nature* (New York: Oxford University Press, 1998), p. 125.

CHAPTER 7. GUARANTEEING DESCENDANTS: THE ROLE OF PARENTAL CARE—TSETSE FLY

1. David L. Denlinger and Jan Ždárek, "Rhythmic pulses of haemolymph pressure associated with partuition and ovulation in the tsetse fly, *Glossina morsitans,*" *Physiological Entomology* 17 (1992): 127–30.

2. Patrick A. Buxton, *The Natural History of Tsetse Flies* (London: H. K. Lewis, 1955), p. xi.

3. Robert E. Snodgrass, *Insects, Their Ways and Means of Living* (New York: Smithsonian Institution Series, 1930), p. 348.

4. Maurice T. James and Robert F. Harwood, *Medical Entomology* (London: Macmillan, 1969), p. 272.

5. Josh Gewolb, "Fake cows fight disease," *Science* 294 (2001): 45.

6. Denlinger and Ždárek, "Rhythmic pulses of haemolymph pressure," pp. 127–30.

7. Buxton, *The Natural History of Tsetse Flies*, p. 106.

8. Ibid., pp. 460–62.

9. Susan Smith, *The Black-Capped Chickadee* (Ithaca, NY: Cornell University Press, 1991), pp. 117–19.

10. Robert L. Metcalf and Robert A. Metcalf, *Destructive and Useful Insects*, 5th ed. (New York: McGraw-Hill, 1993), pp. 20–37.

11. Beth J. Lenoble and David L. Denlinger, "The milk gland of the sheep ked, *Melophagus ovinus*: a comparison with *Glossina*," *Journal of Insect Physiology* 28 (1982): 165–72.

CHAPTER 8. SURVIVING WINTER AS A SLEEPING EGG: EVERGREEN BAGWORM

1. Robert D. Morden and Gilbert P. Waldbauer, "Seasonal and daily emergence patterns of adult *Thyridopteryx ephemeraeformis* (Lepidoptera: Psychidae)," *Entomological News* 82 (1971): 219–24.

2. Tohko Kaufmann, "Observations on the biology and behavior of the evergreen bagworm moth, *Thyridopteryx ephemeraeformis* (Lepidoptera: Psychidae)," *Annals of the Entomological Society of America* 61 (1968): 38–44.

3. David L. Cox and Daniel A. Potter, "Aerial dispersal behavior of larval bagworms, *Thyridopteryx ephemeraeformis* (Lepidoptera: Psychidae)," *Canadian Entomologist* 118 (1986): 525–36.

4. Leonard Haseman, *The Evergreen Bagworm* (Columbia: University of Missouri Agricultural Experimental Station Bulletin 104, 1912), p. 309.

5. Ibid., p. 318.

6. Peter K. Lagoy and Edward M. Barrows, "Larval-sex and host-specific effects on location of attachment sites of last-instar bagworms, *Thyridopteryx ephemeraeformis* (Lepidoptera: Psychidae)," *Proceedings of the Entomological Society of Washington* 91 (1989): 468–72.

7. Gilbert P. Waldbauer and James G. Sternburg, "Cocoons of *Callosamia promethea* (Saturniidae): adaptive significance of differences in mode of attachment to the host tree," *Journal of the Lepidopterists' Society* 36 (1982): 192–99.

8. Morden and Waldbauer, "Seasonal and daily emergence patterns of adult *Thyridopteryx ephemeraeformis*," pp. 219–24.

9. Haseman, *The Evergreen Bagworm*, p. 320.

10. B. A. Leonhardt, J. W. Neal Jr., J. A. Klum, M. Schwarz, et al., "An unusual lepidopteran sex pheromone system in the bagworm moth," *Science* 219 (1983): 314–16.

11. Haseman, *The Evergreen Bagworm*, p. 321.

12. Ibid., pp. 322–23.

13. Robert D. Morden and Gilbert P. Waldbauer, "Diapause and its termination in the psychid moth, *Thyridopteryx ephemeraeformis*," *Entomologia Experimentale et Applicata* 28 (1980): 322–33.

14. Ibid.

15. Gilbert P. Waldbauer and James G. Sternburg, "The bimodal emergence curve of adult *Hyalophora cecropia*: conditions required for the initiation of development by second mode pupae," *Entomologia Experimentale et Applicata* 41 (1986): 315–17.

16. Robert D. Morden and Gilbert P. Waldbauer, "Embryonic development time and spring hatching of *Thyridopteryx ephemeraeformis* (Lepidoptera: Psychidae)," *Entomological News* 82 (1971): 209–17.

17. G. P. Waldbauer, "Phenological adaptation and the polymodal emergence patterns of insects," in *Evolution of Insect Migration and Diapause*, ed. H. Dingle (New York: Springer-Verlag, 1978), pp. 127–44.

18. Khidir W. Hilu and J. M. J. de Wet, "Effects of artificial selection on grain dormancy in *Eleusine* (Gramineae)," *Systematic Botany* 5 (1980): 54–60.

CHAPTER 9. ESCAPING PREDATORS BY DECEPTION: BLACK SWALLOWTAIL BUTTERFLY

1. Paul A. Opler and George O. Krizek, *Butterflies East of the Great Plains* (Baltimore: Johns Hopkins University Press, 1984), p. 45.

2. A. Newnham, "The detailed resemblance of an Indian lepidopterous larva to the excrement of a bird. A similar result obtained in an entirely different way by a Malayan spider," *Transactions of the Entomological Society of London* 282 (1924): xc–xciv.

3. Opler and Krizek, *Butterflies East of the Great Plains,* p. 45.

4. May R. Berenbaum, "Aposematism and mimicry in caterpillars," *Journal of the Lepidopterists' Society* 49 (1995): 386–96.

5. David L. Evans and Gilbert P. Waldbauer, "Behavior of adult and naïve birds when presented with a bumblebee and its mimic," *Zeitschrift für Tierpsychologie* 59 (1982): 247–59.

6. Fritz Müller, "*Ituna* and *Thyridis*; a remarkable case of the mimicry in butterflies" (translated from the German by R. Meldola), *Proceedings of the Entomological Society of London* 27 (1879): xx–xxix.

7. Berenbaum, "Aposematism and mimicry in caterpillars," pp. 386–96.

8. W. Windecker, "*Euchelia (Hypocrita) jacobaeae* L. und das Schutztrachten Problem" (*Euchelia {Hypocrita} jacobaeae* L. and the question of warning coloration), *Zeitschrift für Morphologie und Ökologie der Tiere* 35 (1939): 84–138.

9. Malcolm Edmunds, *Defence in Animals* (Harlow, England: Longman Group, 1974), p. 65.

10. Thomas Eisner and Yvonne C. Meinwald, "Defensive secretion of a caterpillar (*Papilio*)," *Science* 150 (1965): 1733–35.

11. David A. West and Wade N. Hazel, "Natural pupation sites of swallowtail butterflies (Lepidoptera: Papilonidae): *Papilio polyxenes* Fabr., *P. glaucus* L. and *Battus philenor* (L.)," *Ecological Entomology* 4 (1979): 387–92.

12. Wade N. Hazel, S. Ante, and B. Stringfellow, "The evolution of environmentally-cued pupal colour in swallowtail butterflies: natural selection for pupation site and pupal colour," *Ecological Entomology* 23 (1998): 41–44.

13. H. B. D. Kettlewell, "Darwin's missing evidence," *Scientific American* 200 (1959): 48–53.

14. Ibid.

15. Jane Van Zandt Brower, "Experimental studies in some North American butterflies. Part II. *Battus philenor* and *Papilio troilus*, *P. polyxenes* and *P. glaucus*," *Evolution* 12 (1958): 123–36.

16. Lincoln P. Brower, "Ecological chemistry," *Scientific American* 200 (1969): 22–30.

17. Laurence M. Cook, Lincoln P. Brower, and John Alcock, "An attempt to verify mimetic advantage in a Neotropical environment," *Evolution* 23 (1969): 344.

18. Gilbert P. Waldbauer and James G. Sternburg, "Saturniid moths as mimics: an alternative interpretation of attempts to demonstrate mimetic advantage in nature," *Evolution* 29 (1975): 650–58.

19. Gilbert P. Waldbauer, "Aposematism and Batesian mimicry: measuring mimetic advantage in natural habitats," *Evolutionary Biology* 22 (1988): 227–59.

20. James G. Sternburg, Gilbert P. Waldbauer, and Michael R. Jeffords, "Batesian mimicry: selective advantages of color pattern," *Science* 195 (1977): 681–83.

CHAPTER 10. WHY INSECTS ARE SUCH PICKY EATERS: CABBAGE WHITE BUTTERFLY

1. E. Verschaffelt, "The cause determining the selection of food in some herbivorous insects," *Proceedings of the Academy of Science in Amsterdam* 13 (1910): 536–42.

2. Louis M. Schoonhoven, L. M. Tibor Jermy, and J. J. A. van Loon, *Insect-Plant Biology* (London: Chapman and Hall, 1998), pp. 6–8.

3. Niklas Janz, K. Nyblom, and S. Nylin, "Evolutionary dynamics of host-plant specialization: a case study of the tribe Nymphalini," *Evolution* 55 (2001): 783–96.

4. Robert T. Yamamoto and Gottfried S. Fraenkel, "The specificity of the tobacco hornworm, *Protoparce Sexta*, to solanaceous plants," *Annals of the Entomological Society of America* 53 (1960): 503–507.

5. Verschaffelt, "The cause determining the selection of food in some herbivorous insects," pp. 536–42.

6. Asgeir J. Thorsteinson, "The chemotactic responses that determine host specificity in an oligophagous insect. (*Plutella maculipennis* (Curt.) Lepidoptera)," *Canadian Journal of Zoology* 31 (1953): 52–72.

7. Gottfried S. Fraenkel, "The raison d'être of secondary plant substances," *Science* 129 (1959): 1466–70.

8. Andreas Ratska, H. Vogel, D. J. Kliebenstein, et al., "Disarming the mustard oil bomb," *Proceedings of the National Academy of Sciences* 99 (2002): 11223–28.

9. Fraenkel, "The raison d'être of secondary plant substances," pp. 1466–70.

10. Ibid.

11. Gilbert P. Waldbauer and Gottfried S. Fraenkel, "Feeding on normally rejected plants by larvae of the tobacco hornworm, *Protoparce sexta* (Lepidoptera: Sphingidae)," *Annals of the Entomological Society of America* 54 (1961): 477–85.

12. Paul R. Ehrlich and Peter H. Raven, "Butterflies and plants: a study in coevolution," *Evolution* 18 (1965): 586–608.

13. G. E. Hutchinson, *The Ecological Theater and the Evolutionary Play* (New Haven, CT: Yale University Press, 1965), title page.

14. Arthur E. Weis and May R. Berenbaum, "Herbivorous insects and green plants," in *Plant-Animal Interaction*, ed. W. G. Abrahamson (New York: McGraw-Hill, 1989), pp. 123–62.

15. Ehrlich and Raven, "Butterflies and plants: a study in coevolution," pp. 586–608.

16. Scott R. Smedley, F. C. Schroeder, D. B. Weibel, et al., "Mayolenes: labile defensive lipids from the glandular hairs of a caterpillar (*Pieris rapae*)," *Proceedings of the National Academy of Sciences* 99 (2002): 6822–27.

CHAPTER 11. "NUTRITIONAL WISDOM": CORN EARWORM

1. R. A. Blanchard and W. A. Douglas, *The Corn Earworm as an Enemy of Field Corn in the Eastern States* (Washington, DC: USDA Farmers' Bulletin 1651, 1953), pp. 1–15.

2. Philip S. Callahan, "Behavior of the imago of the corn earworm, *Heliothis zea* (Boddie), with special reference to emergence and reproduction," *Annals of the Entomological Society of America* 51 (1958): 271–83.

3. Robert L. Metcalf and Robert A. Metcalf, *Destructive and Useful Insects*, 5th ed. (New York: McGraw-Hill, 1993), p. 9.34.

4. L. G. Monteith, "Host preferences of *Drino bohemica* Mesh. (Diptera: Tachinidae), with particular reference to olfactory responses," *Canadian Entomologist* 87 (1955): 509–30.

5. Ted C. J. Turlings, J. H. Loughrin, P. J. McCall, et al., "How caterpillar-damaged plants protect themselves by attracting parasitic wasps," *Proceedings of the National Academy of Sciences* 92 (1995): 4169–74.

6. Gary F. McCracken and John K. Westbrook, "Bat patrol," *National Geographic* 201 (April 2002): 114–23.

7. Ibid.

8. Jane E. Yack and James H. Fullard, "Ultrasonic hearing in nocturnal butterflies," *Nature* 403 (2000): 265–66.

9. Gilbert P. Waldbauer and Anoop K. Bhattacharya, "Self-selection of an optimum diet from a mixture of wheat fractions by the larvae of *Tribolium confusum*," *Journal of Insect Physiology* 19 (1973): 407–18.

10. Erma S. Vanderzant, "Dietary requirements of the bollworm, *Heliothis zea* (Lepidoptera: Noctudiae), for lipids, choline, and inositol and the effect of fats and fatty acids on the composition of the body fat," *Annals of the Entomological Society of America* 61 (1968): 120–25.

11. Gilbert P. Waldbauer and Stanley Friedman, "Self-selection of optimal diets by insects," *Annual Review of Entomology* 36 (1991): 43–63.

12. Konrad Lorenz, *Evolution and the Modification of Behavior* (Chicago: University of Chicago Press, 1965), pp. 14–15.

13. Robert W. Thacker, "Avoidance of recently eaten foods by land hermit crabs, *Coenobita compressus*," *Animal Behavior* 55 (1998): 485–96.

14. Tohko Kaufmann, "Biological studies on some Bavarian Acridoidea (Orthoptera), with special reference to their feeding habits," *Annals of the Entomological Society of America* 58 (1965): 791–801.

15. Susan F. Reichert and Joel M. Harp, "Nutritional ecology of spiders," in *Nutritional Ecology of Insects, Mites, Spiders, and Related Invertebrates*, ed. F. Slansky Jr. and J. G. Rodriguez (New York: Wiley, 1987), pp. 645–72.

CHAPTER 12. INVADERS FROM ABROAD: GYPSY MOTH

1. A. F. Burgess, "Suppression of the gipsy [sic] and brown-tail moths and its value to states not infested," *Yearbook of the Department of Agriculture* (Washington, DC, 1916), pp. 1–2.

2. Edward H. Forbush and Charles H. Fernald, *The Gypsy Moth* (Boston: Massachusetts State Board of Agriculture, 1896), p. 11.

3. D. Barry Lyons and Andrew M. Liebhold, "Spatial distribution and hatch times of egg masses of gypsy moth (Lepidoptera: Lymantriidae)," *Environmental Entomology* 21 (1992): 354–58.

4. W. E. Britton, *The Gypsy Moth* (New Haven: Connecticut Agricultural Experiment Station Bulletin 375, 1935), pp. 625–47.

5. Michael H. Gerardi and James K. Grimm, *The History, Biology, Damage, and Control of the Gypsy Moth, Porthetria dispar* (L.) (Cranbury, NJ: Associated University Presses, 1979), p. 287.

6. Britton, *The Gypsy Moth,* p. 631.

7. Gerardi and Grimm, *The History, Biology, Damage, and Control of the Gypsy Moth,* p. 168.

8. Ibid., p. 169.

9. Burgess, "Suppression of the gipsy [sic] and brown-tail moths," p. 5.

10. Ibid., p. 2.

11. Ibid.

12. Britton, *The Gypsy Moth*, p. 638.

13. Marjorie A. Hoy, "Parasitoids and predators in management of arthropod pests," in *Introduction to Insect Pest Management,* 3rd ed., ed. R. L. Metcalf and W. H. Luckmann (New York: John Wiley, 1994), pp. 160–61.

14. Gerardi and Grimm, *The History, Biology, Damage, and Control of the Gypsy Moth,* p. 134.

15. Curtis W. Sabrosky and Paul H. Arnaud Jr., "Family Tachinidae," in *A Catalog of the Diptera of America North of Mexico*, ed. A. Stone, C. W. Sabrosky, W. W. Wirth, et al. (Washington, DC: US Government Printing Office, 1965), p. 1040.

16. George H. Boettner, J. S. Elkinton, and C. J. Boettner, "Effects of a biological control introduction on three nontarget native species of saturniid moths," *Conservation Biology* 14 (2000): 1798–1806.

17. David Pimentel, L. Lach, and R. Zuniga, "Environmental and economic costs of nonindigenous species in the United States," *BioScience* 50 (2000): 53–65.

CHAPTER 13. AN AMERICAN SAVES THE FRENCH WINE INDUSTRY: GRAPE PHYLLOXERA

1. George Ordish, *The Great Wine Blight* (New York: Charles Scribner's Sons, 1972), p. 8.

2. Edward H. Smith, "The grape phylloxera, a celebration of its own," *American Entomologist* 38 (1992): 212–21.

3. Ordish, *The Great Wine Blight*, p. 140.

4. Smith, "The grape phylloxera, a celebration of its own," pp. 212–21.

5. John H. Comstock, *An Introduction to Entomology* (Ithaca, NY: Comstock Publishing Co., 1950), pp. 433–37.

6. Smith, "The grape phylloxera, a celebration of its own," pp. 212–21.

7. Jeffrey Granett, M. A. Walker, L Kocis, et al., "Biology and management of grape phylloxera," *Annual Review of Entomology* 46 (2001): 387–412.

8. Gerald A. Rosenthal and May R. Berenbaum, eds., *Herbivores, Their Interactions with Secondary Plant Metabolites*, vol. 2 (San Diego: Academic Press, 1992). Refer to whole book.

9. Gerald A. Rosenthal, C. G. Hughes, and D. H. Janzen, "Canavanine, a dietary nitrogen source for the seed predator *Caryedes brasiliensis* (Bruchidae)," *Science* 217 (1982): 353–55.

10. G. Wayne Ivie, D. L. Bull, R. C. Beier, et al., "Metabolic detoxification: mechanism of insect resistance to plant psoralens," *Science* 221 (1983): 374–76.

11. May Berenbaum, "Patterns of furanocoumarin distribution and insect herbivory in the Umbelliferae: plant chemistry and community structure," *Ecology* 62 (1981): 1254–66.

CHAPTER 14. AN INSECTICIDE "CREATES" NEW PESTS: CODLING MOTH

1. Mark L. Winston, *Nature Wars: People vs. Pests* (Cambridge, MA: Harvard University Press, 1997), p. 82.

2. Robert L. Metcalf and Robert A. Metcalf, *Destructive and Useful Insects*, 5th ed. (New York: McGraw-Hill, 1993), pp. 15.45–15.49.

3. Winston, *Nature Wars,* pp. 83–84.

4. Thomas R. Dunlap, *DDT, Scientists, Citizens, and Public Policy* (Princeton, NJ: Princeton University Press, 1982), p. 43.

5. Ibid., p. 50.

6. Ibid., p. 54.

7. E. H. Glass and P. J. Chapman, *The Red-banded Leaf Roller and Its Control,* Bulletin 755 (Geneva, NY: New York State Agricultural Experiment Station, 1952), p. 4.

8. C. R. Cutright, *Grower Control of the Codling Moth* (Wooster: Ohio Agricultural Experiment Station Research Bulletin 720, 1953), p. 35.

9. Metcalf and Metcalf, *Destructive and Useful Insects*, p. 7.48.

10. Glass and Chapman, *The Red-banded Leaf Roller and Its Control*, p. 4.

11. R. H. Meyer, "Management of phytophagous and predatory mites in Illinois orchards," *Environmental Entomology* 3 (1974): 333–40.

CHAPTER 15. FROM LOW- TO HIGH-TECH CONTROLS: EUROPEAN CORN BORER

1. W. G. Bradley, "The European corn borer," in *Yearbook of Agriculture* (Washington, DC: United States Department of Agriculture, 1952), p. 614.

2. D. J. Caffrey and L. H. Worthley, *A Progress Report on the Investigations of the European Corn Borer*, Department Bulletin No. 1476 (Washington, DC: United States Department of Agriculture, 1929), pp. 15–24.

3. D. G. R. McLeod and S. D. Beck, "Photoperiodic termination of diapause in an insect," *Biological Bulletin* 124 (1963): 84–96.

4. H. C. Chiang, "Overwintering corn borer, *Ostrinia nubilalis*, larvae in storage cribs," *Journal of Economic Entomology* 57 (1964): 666–69.

5. Dean Barry, D. Alfaro, and L. L. Darrah, "Relation of European corn borer (Lepidoptera: Pyralidae) leaf-feeding resistance and DIMBOA content in maize," *Environmental Entomology* 23 (1994): 177–82.

6. William H. Luckmann and George C. Decker, "A corn plant maturity index for use in European corn borer ecological and control investigations," *Journal of Economic Entomology* 45 (1952): 226–32.

7. Stanley D. Beck and Edward E. Smissman, "The European corn borer, *Pyrausta nubilalis*, and its principal host plant. IX. Biological activity of chemical analogs of core resistance factor A (6-methoxybenzoxazilone)," *Annals of the Entomological Society of America* 54 (1961): 52–61.

8. Luckmann and Decker, "A corn plant maturity index for use in European corn borer ecological and control investigations," pp. 226–32.

9. William H. Luckmann and Howard B. Petty, *Controlling Corn Borers in Field Corn with Insecticides*, Circular 78 (Champaign, IL: College of Agriculture Extension Service in Agriculture, and Home Economics, 1957), pp. 1–8.

10. John E. Losey, L. S. Rayor, and M. E. Carter, "Transgenic pollen harms monarch larvae," *Nature* 399 (1999): 214.

11. Arthur R. Zangerl, D. McKenna, C. L. Wraight, et al., "Effects of exposure to event 176 *Bacillus thuringiensis* corn pollen on monarch and swallowtail caterpillars under field conditions," *Proceedings of the National Academy of Sciences* 98 (2001): 11908–12.

12. Ibid.

13. May Berenbaum, ed., "B_t Corn and Monarch Butterflies," *Proceedings of the National Academy of Sciences* 98 (2001): 11908–42.

14. Leonard I. Wassenaar and Keith A. Hobson, "Natal origins of migratory monarch butterflies at wintering colonies in Mexico: new isotopic evidence," *Proceedings of the National Academy of Sciences* 95 (1998): 15436–39.

15. Rachel Carson, *Silent Spring* (New York: Houghton Mifflin, 1962). Refer to entire book.

CHAPTER 16. THE DEMISE OF DDT: JAPANESE BEETLE

1. E. Dwight Sanderson, *Insect Pests of Farm, Garden, and Orchard* (New York: John Wiley and Sons, 1931), p. 320.

2. William H. Luckmann and George C. Decker, "A 5-year report of observations in the Japanese beetle control area at Sheldon, Illinois," *Journal of Economic Entomology* 53 (1960): 821–27.

3. Thomas G. Scott, Y. L. Willis, and J. A. Ellis, "Some effects of a field application of dieldrin on wildlife," *Journal of Wildlife Management* 23 (1959): 409–27.

4. Rachel Carson, *Silent Spring* (New York: Houghton Mifflin, 1962), p. 92.

5. Ibid., p. 13.

6. Thomas R. Dunlap, *DDT* (Princeton, NJ: Princeton University Press, 1981), p. 139.

7. Paul R. Ehrlich and Anne H. Ehrlich, *Population, Resources, Environment*, 2nd ed. (San Francisco: W. H. Freeman, 1972), p. 390.

8. Dunlap, *DDT*, p. 99.

9. Carson, *Silent Spring*, p. 22.

10. Mary L. Flint and Robert van den Bosch, *Introduction to Integrated Pest Management* (New York: Plenum Press, 1981), p. 6.

11. Walter E. Fleming, *Controlling the Japanese Beetle*, Farmer's Bulletin 2004 (Washington, DC: US Department of Agriculture, 1955), p. 12.

CHAPTER 17. THEIR PASSING FROM THE AGRICULTURAL SCENE: CHINCH BUG

1. F. A. Fenton and C. F. Stiles, *Chinch Bug Control for Oklahoma Farms* (Stillwater: Oklahoma Agricultural and Mechanical College, 1940), p. 18.

2. C. M. Packard, P. Luginbill Sr., and C. Benton, *How to Fight the Chinch Bug,* Farmer's Bulletin 1780 (Washington, DC: US Department of Agriculture, 1937), p. 3.

3. *Chinch Bugs: How to Control Them,* Leaflet 364 (Washington, DC: US Department of Agriculture, 1954), p. 5.

4. W. L. Burlison and W. P. Flint, *Fight the Chinch-Bug with Crops,* Experiment Circular 268 (Urbana: University of Illinois Agricultural College, 1923), p. 3.

5. Robert L. Metcalf and Robert A. Metcalf, *Destructive and Useful Insects*, 5th ed. (New York: McGraw-Hill, 1993), p. 11.32.

6. Gilbert P. Waldbauer, "Damage to soybean seeds by South American stink bugs," *Anais Da Sociedade Entomológica Do Brasil* 6 (1977): 224–29.

7. Norman L. Marston, G. D. Thomas, C. M. Ignoffo, et al., "Seasonal cycles of soybean arthropods in Missouri: effect of pesticidal and cultural practices," *Environmental Entomology* 8 (1979): 165–73.

8. Larry Zuckerman, *The Potato: How the Humble Spud Rescued the Western World* (New York: North Point Press, 1998). Refer to entire book.

9. W. Lu and J. Lazell, "The voyage of the beetle," *Natural History* 105 (1996): 36–39.

10. Peter W. Price, *Insect Ecology*, 3rd ed. (New York: John Wiley and Sons, 1997), p. 103.

CHAPTER 18. SYNCHRONY WITH THE SEASONS: HESSIAN FLY

1. W. B. Cartwright and E. T. Jones, *The Hessian Fly,* Farmer's Bulletin 1627 (Washington, DC: US Department of Agriculture, 1953), pp. 1–9.

2. Robert A. Metcalf and Robert L. Metcalf, *Destructive and Useful Insects*, 5th ed. (New York: McGraw-Hill, 1993), pp. 10.8–10.12.

3. Charles G. Helm, Michael R. Jeffords, Susan L. Post, et al., "Spring feeding activity of overwintered bean leaf beetles (Coleoptera: Chrysomelidae) on nonleguminous hosts," *Environmental Entomology* 12 (1983): 321–22.

4. Gilbert P. Waldbauer and Marcos Kogan, "Bean leaf beetle: phenological relationship with soybean in Illinois," *Environmental Entomology* 5 (1976): 35–44.

5. Carol K. Augspurger, "Reproductive synchrony of a tropical shrub: experimental studies on effects of pollinators and seed predators on *Hybanthus prunifolius* (Violaceae)," *Ecology* 62 (1981): 775–78.

6. Gilbert P. Waldbauer and Joseph K. Sheldon, "Phenological relationship of some aculeate Hymenoptera, their dipteran mimics, and insectivorous birds," *Evolution* 25 (1971): 371–82.

7. Miriam Rothschild, "An extension of Dr. Lincoln Brower's theory on bird predation and food specificity, together with some observations on bird memory in relation to aposematic colour patterns," *Entomologist* (London) (1964): 73–78.

8. David Tilman, "Cherries, ants, and tent caterpillars: timing of nectar production in relation to susceptibility of caterpillars to ant predation," *Ecology* 59 (1978): 686–92.

9. Paul Feeny, "Seasonal changes in oak leaf tannins and nutrients as a cause of spring feeding by winter moth caterpillars," *Ecology* 51 (1970): 565–81.

10. John C. Schneider, "The role of parthenogenesis and female aptery in microgeographic ecological adaptation in the fall cankerworm, *Alsophila pometary* Harris (Lepidoptera: Geometridae)," *Ecology* 61 (1980): 1082–90.

CHAPTER 19. AN INSECT TO CONTROL ANOTHER INSECT: COTTONY CUSHION SCALE

1. Paul DeBach, *Biological Control by Natural Enemies* (London: Cambridge University Press, 1974), pp. 31–38.

2. Ibid., p. 99.

3. Ibid., p. 33.

4. Michael Kosztarab, "Everything unique or unusual about scale insects (Homoptera: Coccoidae)," *Bulletin of the Entomological Society of America* 33 (1987): 215–20.

5. DeBach, *Biological Control by Natural Enemies,* p. 34.

6. Albert Koebele, *Report of a Trip to Australia to Investigate the Natural Enemies of the Fluted Scale*, Bulletin 21 (Washington, DC: US Department of Agriculture Division of Entomology, 1890), p. 5.

7. Ibid., p. 12.

8. DeBach, *Biological Control by Natural Enemies*, p. 97.

9. Ibid., p. 92.

10. J. D. Tothill, T. H. C. Taylor, and R. W. Paine, *The Coconut Moth in Fiji* (London: Imperial Bureau of Entomology, 1930), pp. 14–17, 33.

11. Donald Naflus, "Biological control of *Penecillaria jocosatrix* (Lepidoptera: Noctuidae) on mango on Guam with notes on the biology of its parasitoids," *Environmental Entomology* 20 (1991): 1725–31.

12. J. K. Holloway, "Projects in biological control of weeds," in *Biological Control of Insect Pests and Weeds*, ed. P. DeBach (New York: Reinhold, 1964), pp. 652–56.

13. D. F. Waterhouse, "The biological control of dung," *Scientific American* 230, no. 4 (1974): 100–109.

14. Francis G. Howarth, "Environmental impacts of classical biological control," *Annual Review of Entomology* 36 (1991): 485–509.

15. E. S. Krasfur, T. J. Kring, J. C. Miller, et al., "Gene flow in the exotic colonizing ladybeetle *Harmonia axyridis* in North America," *Biological Control* 8 (1997): 207–14.

CHAPTER 20. EXTERMINATION BY SUBVERTING THE SEX ACT: SCREWWORM FLY

1. Harold Oldroyd, *The Natural History of Flies* (New York: W. W. Norton, 1964), p. 218.

2. W. E. Dove, "Myiasis of man," *Journal of Economic Entomology* 30 (1937): 29–39.

3. Edward Knipling, "Possibilities of insect control or eradication through the use of sexually sterile males," *Journal of Economic Entomology* 48 (1955): 459–62.

4. A. H. Baumhover, A. J. Graham, B. A. Bitter, et al., "Screw-worm control through release of sterilized flies," *Journal of Economic Entomology* 48 (1955): 462–66.

5. Knipling, "Possibilities of insect control or eradication," pp. 459–62.

6. Baumhover et al., "Screw-worm control through release of sterilized flies," pp. 462–66.

7. L. F. Steiner, W. C. Mitchell, E. J. Harris, et al., "Oriental fruit fly eradication by male annihilation," *Journal of Economic Entomology* 58 (1965): 961–64.

8. F. M. Howlett, "The effect of oil of citronella on two species of *Dacus*," *Transactions of the Entomological Society of London, Part 2* (1912): 412–18.

EPILOGUE

1. Paul A. Colinvaux, *Why Big Fierce Animals Are Rare* (Princeton, NJ: Princeton University Press, 1978), p. 8.

2. Elizabeth Bernays and Michelle Graham, "On the evolution of host specificity in phytophagous arthropods," *Ecology* 69 (1988): 886–92.

3. Gilbert P. Waldbauer and Stanley Friedman, "Self-selection of optimal diets by insects," *Annual Review of Entomology* 36 (1991): 43–63.

4. Edward O. Wilson, *Sociobiology* (Cambridge, MA: Harvard University Press, 1975), p. 208.

5. Louis M. Roth and Edwin R. Willis, *The Biotic Associations of Cockroaches* (Washington, DC: Smithsonian Institute, 1960), p. 329.

6. Robert L. Metcalf and Robert A. Metcalf, *Destructive and Useful Insects*, 5th ed. (New York: McGraw-Hill, 1993), pp. 1.41–1.42.

7. Rachel Carson, *Silent Spring* (New York: Houghton Mifflin, 1962), pp. 97–119.

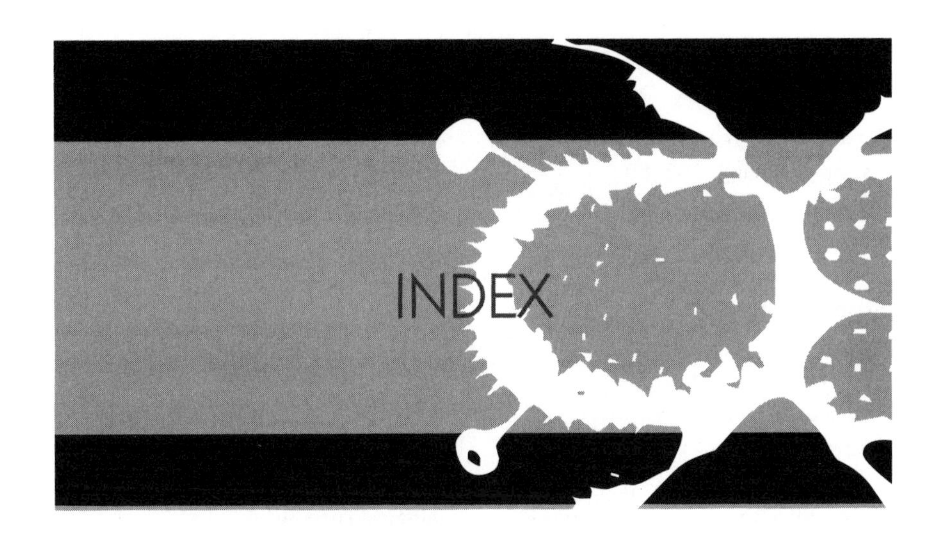

acaricide, 191, 193
Acyrthosiphon pisum. See pea aphid
adults, 19
  cabbage white butterfly, 131
  chinch bug, 220
  corn earworm moth, 145
  corn rootworms, 62
  Drosophila, 51
  European corn borer, 196
  evergreen bagworm, 102, 104
  grape phylloxera, 173
  Hessian fly, 235
  house fly, 42–43
  Japanese beetle, 208
  mosquitoes, 22
  tsetse fly, 96
Aedes aegypti. See yellow-fever mosquitoes
Aedes tormentor. See mosquitoes
Aedes vexans. See mosquitoes
AIDS and mosquitoes, 35
"alarm scents," 146
Alcock, John, 126
aldicarb, 188
aldrin, 214
allantoin, 255
allele, 54, 88–89
Amherst College, 124
Anais Da Sociedade Entomológica Do Brasil, 226
Anopheles. See malaria mosquitoes
ants
  and caterpillars, 119–20, 140–41
  killing other insects, 240
  and mosquitoes, 22
aphids, 46, 81–89, 252, 269
  grape phylloxera, 171–82
apple maggot, 70–71, 73, 266
  fruit selection by, 76–78
  See also fruit fly, Tephritidae
apples, 73, 183–85, 188, 190–91
army caterpillars, 146

Arnaud, Paul, Jr., 167
arsenic poisoning, 188–89
Artona moth, 248
Ashburner, M., 51
Asian ladybird beetles, 252
Asian long-horned beetle, 170
Asian scale insect, 245
Asian tiger mosquito, 169–70
Augspurger, Carol, 238
Australian ladybird beetle, 242
Australian scarabs, 250–51
automobiles and the spread of gypsy moths, 162

Bacillus thuringiensis, 166, 169, 194, 201, 202, 203
  Max 454, 203–205
Bactrocera dorsalis. *See* oriental fruit fly
bagworm, 93, 99–111
balancers, 47
ballooning, 101
bark beetles, 193, 212
barrier to chinch bugs, 219, 222–24
Barrows, Edward, 103
Barry, Dean, 199
Bates, Henry, 118
Batesian mimicry. *See* defenses of insects
bats, 147–48, 186, 268
  insect's protection against, 148, 268
Baumhover, A. H., 258, 259
bean leaf beetles, 131, 227–28, 237–38
Beck, Stanley, 196, 199
beetles, 131
  Asian ladybird beetles, 252
  Asian long-horned beetle, 170
  Australian ladybird beetle, 242
  burying beetles, 270
  corn rootworms, 61–68
  elm leaf beetles, 131
  flour beetle, 149–50
  Japanese beetle, 169, 207–18
  scarab beetles, 250–51
Belcher, William (Mrs.), 156–57
Berenbaum, May, 117, 139, 140, 180, 181, 203, 204
Berlocher, Stewart, 71, 73
Bhattacharya, Anoop, 149
biochemical defenses of plants. *See* plant and insect interaction
biodegrading of insecticides, 209
biological controls
  adverse side effects, 251–52
  of insects, 166–68, 242, 246, 247–51
  male annihilation technique, 261–64
  of plants, 231–32
  sterile male technique, 258–61
biotechnology, 201
bird droppings, mimicry of, 117, 208n2, 268
birds, 30, 81, 122, 186
  affected by insecticides, 40, 209, 210, 212, 214, 215
  avoidance of toxic insects, 118, 119, 123–24, 239–40
  larger-than-life artificial eggs, 77
  parental care, 96–97
  and sychrony, 242
  and West Nile Fever, 27–29
Bishopp, F. C., 43
black cherry trees and ants, 240
black peppered moth, 121–22, 266
black swallowtail butterfly, 113–28, 181, 182, 239, 268
  impact of GMO on, 202–205
  parsleyworm, 16, 181
Blanchard, R. A., 144
Blepharella, 249

Blepharoneura, 71
Blissus leucopterus. *See* chinch bug
blood as food, 22, 23, 25, 92, 95, 254
blow fly, 253, 254–55
bluebottle flies, 253
bluffing. *See* mimicry to avoid predators
Boettner, George, 168
Boller, E. F., 74
boll weevil, 194
Bradley, William, 195
Branson, Terry, 63
breeding sites for mosquitoes, 18
Britton, W. E., 158, 159, 166
Brower, Jane Van Zandt, 123
Brower, Lincoln, 124, 125–26
Brown, Anthony, 40
B_t. *See Bacillus thuringiensis*
buffalo bur, 230–31
Bureau of Chemistry of the USDA. *See* US Food and Drug Administration
Burgess, A. F., 156, 162, 163, 164
Burlison, W. L., 224
burying beetles, 270
Bushland, Raymond, 260
butterflies, 117, 122, 123
  black swallowtail, 16, 113–28, 181, 182, 202–205, 239, 268
  cabbage white, 129–41
  difference from flies, 38
  Masculina arion, 251
  monarch, 114, 118–19, 124, 131, 133, 202–05
  pipevine swallowtail, 122–23, 125, 127
  spice bush swallowtail, 122, 123
  viceroy, 114, 124
Buxton, Patrick, 92, 96

cabbage white butterfly, 129–41
Cactoblastis moth, 250
cactus, prickly pear, 249–50
Caffrey, D. J., 196
California Citrus Growers' Association, 246
Callahan, Philip, 145
Calliphoridae, 254
camouflage as a defense. *See* defenses of insects
Campbell, Walter G., 189
cannibalism of corn earworm moth, 143
carbaryl, 166
Carey, James, 72
Carlton University, 148
carrion-feeding, 253, 254–55
Carson, Rachel, 205, 210–11, 215–16, 217, 272
Cartwright, W. B., 234
Caryedes brasiliensis. *See* seed weevil
caterpillar lime, 164
caterpillars. *See* larvae
cattle
  "fake cows," 95, 272
  and screwworms, 256
  and tsetse fly, 92, 95
cecropia moth, 108–109, 110, 168
Central High School (Bridgeport, CT), 37
Ceratitis capitata. *See* Mediterranean fruit fly
Chapman, P. J., 190, 191
chemical defense of plants. *See* plant and insect interaction
chemicals to avoid predators. *See* defenses of insects
Chiang, H. C., 66, 198
chinch bug, 219–32
  use of barriers against, 219, 222–24
chlordane, 214

chlorinated hydrocarbon insecticides. *See* DDT; insecticides
chloroquine, 33
chromosomes, 53–54, 56–57, 58–59
chrysalis. *See* pupae
Chrysolina. *See* leaf beetle
cinnabar moth, 119
circumventing crop rotation. *See* crop rotation
citronella, 262
Clark, Ronald W., 57, 58
clean plowing, 198
clones, 84, 85–89
cochineal scale, 245
Cochliomyia hominovorax. *See* screwworm fly
cockroaches, 151, 270
cocoons. *See* pupae
codling moth, 183–94
  sterile male technique, 261
coevolution, 139–40, 174
  See also evolution
Colinvaux, Paul, 266
Colorado potato beetle, 131, 170, 229–31
coloration to avoid predators. *See* defenses of insects
Columbia University, 55
complete metamorphosis. *See* metamorphosis
Compsilura concinnata. *See* Eurasian fly
Comstock, John Henry, 173
Congo floor maggots, 254
Connecticut Agricultural Experiment Station, 158
contact poisons, 191
Cook, Lincoln, 126
copra industry, 248
corn earworm moth, 143–54, 267
Cornell University, 169, 172, 173, 202
corn plants, 197–98
  adverse effect of GMO, 202–205
  tassel ratio, 199–201
Corn Pollen and Monarch Butterflies (Berenbaum), 204
corn rootworms, 61–68
Corn/Soybeans Study Team, 225
Correns, Carl, 54
cotton bollworm. *See* corn earworm moth
cottony cushion scale, 166, 193, 242, 243–52
Cox, David, 101
crawlers, 245
creosote, 163, 223
cricket, 93
crop rotation, 59, 61–62
  circumventing, 65–68
crossing over (genetics), 58–59
cruciferae. *See* mustard family of plants
Culex perfidiosus. *See* mosquitoes
Culex pipiens. *See* house mosquito
Culex quinquefasciatus. *See* mosquitoes
Culiseta. *See* mosquitoes
Cunningham, Roy, 262–63
Cutright, C. R., 190
cyanide in apples, 184
cyclone burner, 163–64
Cydia pomonella. *See* codling moth

Dacus zonatus. *See* peach fruit fly
Daktulosphaira vitifoliae. *See* grape phylloxera
D'Antonio, Michael, 18, 27, 29, 32
Darwin, Charles, 41, 48, 52–53
  natural selection, 56
DDD, 190, 212
DDE, 40, 213

DDT, 187, 212, 213, 248, 271
  ban of use, 205, 209, 214
  causing new insect problems, 190–92, 193
  dangers of, 207–18
  residual traces, 32, 209, 215
  resistance to, 33, 40–41, 45, 59, 266
  side effects, 188
  See also insecticides
DeBach, Paul, 243, 248
debriding wounds using blow flies, 255
deception to avoid predators. *See* defenses of insects
Decker, George, 199–200, 208–209
decomposition and blow flies, 253–54
Deer Lodge National Forest (Montana), 37
Defence in Animals (Edmunds), 119
defense against predators. *See* predators, avoiding
defenses of insects, 113–28, 268
  camouflage as a defense, 41–42, 117, 120–22, 132, 183, 268
  chemical defense, 119–20, 140–41
  mimicry as defense, 113–14, 122–23, 239
    Batesian mimicry, 118–19, 122, 123–27
    Müllerian mimicry, 118, 124
    resembling bird droppings, 117, 268, 280n2
  noxiousness as defense, 119, 122–23, 133, 240
  physiological defenses, 181
  warning coloration in defense, 117–18, 122–23, 125–26, 133
defenses of plants. *See* plant and insect interaction
defoliation by gypsy moths, 156–59, 165, 167
Denlinger, David, 91, 95, 97
derris, 209
detoxification and insects, 181–82, 184, 267
De Vries, Hugo, 54
Diabrotica barberi. *See* northern corn rootworm
Diabrotica virgifera. *See* western corn rootworm
diamond-back moth, 136–38
Diana fritillary, 122
diapause, 66, 98, 106–107, 108–11, 267–68
  apple maggot, 71, 74
  black swallowtail butterfly, 115
  cecropia moth, 108–109
  chinch bug, 220, 222
  codling moth, 185
  corn rootworms, 62, 66
  European corn borer, 196
  European red mite, 192
  evergreen bagworm, 106–107
  grape phylloxera, 174
  gypsy moth, 159
  hawthorn maggot, 71
  Japanese beetle, 208
  mosquitoes, 25
  silkworm, 107–08
  in summer, 107
Díaz, Francisco, 78–79
dieldrin. *See* insecticides
diet, balanced. *See* nutrient self-selection
DIMBOA, 198–200, 201
Diptera. *See* flies
diseases, 18, 34–35, 38
  elephantiasis, 35
  encephalitis, 34
  filariasis, 26, 35

malaria, 18, 25, 30–34, 271
nagana, 92, 96
sleeping sickness, 92, 93, 96
West Nile Fever, 18, 27–29
yellow fever, 18, 25–27
disparlure, 161, 166
dispersal of insects, 38, 43, 49–50, 161–62, 163–65, 208–209, 231
divergence. *See* evolutionary radiation
Dixon, A. F. G., 87
DNA, 55, 59, 147–48, 181, 201
dominant allele, 54
Doppler radar, 147–48
dormancies, 111
See also diapause
double helix. *See* DNA
Douglas, W. A., 144
Dove, W. E., 256
Downes, William, 17
Drosophila, 49–60
advantages as lab animal, 55–56
differences from Teprhitidae, 69
white-eyed male mutant, 57–58
Dunlap, Thomas, 39, 188, 189, 213–14, 215
Dutch elm disease, 193, 212

eastern tent caterpillar, 240
echolocation, 148, 268
ecological disasters, 271–72
defoliation by gypsy moths, 156–59
use of insecticides, 189–90, 193, 209, 210
Ecological Theater and the Evolutionary Play, The (Hutchinson), 139
ecosystem balance, 167, 189–90, 224, 265
Edmunds, Malcolm, 119
egg, 19
egg diapause, 108
See also diapause
eggs
apple maggot, 77–78
black swallowtail butterfly, 114
cabbage white butterfly, 131
chinch bug, 220–21, 222
codling moth, 183
corn earworm moth, 144
corn rootworms, 62, 64–65, 66, 67
Drosophila, 51
European corn borer, 196
evergreen bagworm, 105
fall cankerworm, 242
grape phylloxera, 173, 174
gypsy moth, 159, 162, 163–64
Hessian fly, 234
house fly, 43–44
Japanese beetle, 208
mosquitoes, 24–25
protection from predators, 116, 117
screwworm fly, 255
synchronization of hatching, 110
tsetse fly, 95
eggshells and DDT, 213–14
Ehrlich, Anne, 214
Ehrlich, Paul, 139, 214
Eisner, Thomas, 120, 167
elephantiasis, 35
elm leaf beetles, 131
encephalitis, 34
endrin, 210, 214
Entomological Society of America, 176
Entomological Society of London, 22
Entomology and Pest Management (Pedigo), 14
environmental disasters. *See* ecological disasters
enzymes in mustard family, 136–38

eradication of insects, 272
gypsy moth, 163–68
male annihilation technique, 261–63
Mediterranean fruit fly, 72–73
mosquitoes, 32–33
sterile male technique, 258–61
Euphranta toxoneura, 72
Euplectrus, 249
Eurasian fly, 167–68
European corn borer, 169, 195–205
European red mite, 190–91, 192
Evans, David, 118
evergreen bagworm, 99–111
evolution, 35–36, 41–42, 180
apple maggot, 73
examples of, 265–66
food choices of insects, 132, 139
and genetics, 55
house fly, 45
mouthparts, 46
retrogressive evolution, 244
See also neo-Darwinian synthesis
evolutionary radiation, 49–50, 139
exported insects. *See* invading insects
extended diapause, 66
See also diapause
external fertilization, 44
eyes
compound, 48
white-eyed male *Drosophila*, 57–58

facultative diapause, 107
See also diapause
"fake cows," 95, 272
falciparum. *See* malaria
fallacis mite, 192
fall armyworm, 182
fall cankerworm, 241–42
false tobacco budworm. *See* corn earworm moth
Farmer's Bulletin. *See* US Department of Agriculture
Farquharson, C. O., 22
Feeny, Paul, 241
females
black swallowtail butterfly, 114, 122–23
cottony cushion scale, 244–45
Drosophila, 50–51
European corn borer, 196
evergreen bagworm, 102, 104–106
grape phylloxera, 173–74
gypsy moth, 159–61
house fly, 43–44, 45
Japanese beetle, 208
mosquitoes, 22–23, 24
promethea, 127
screwworm fly, 255, 258
Fenton, F. A., 222
Fernald, Charles, 156–57
fertilization, 44
Fight the Chinch-Bug with Crops, 224
filariasis, 26, 35
firefly, 93
"flax seed stage," 235
Fleming, Walter, 217
Flemming, Walther, 54
flesh-eating insects, 253–64
flies, 253
apple maggot, 70–71, 73, 76–78, 266
Drosophila, 49–60
Eurasian fly, 167–68
flower flies, 76, 239
hawthorn maggot, 70–71
Hessian fly, 169, 233–42, 269
house fly, 37–48
hover flies, 93

human bot fly, 254
oriental fruit fly, 74, 261–64
peach fruit fly, 262
screwworm fly, 253–64
Tephritidae, 49, 69–79
tsetse fly, 91–98, 269
flight of house fly, 47
Flint, Mary, 216
Flint, W. P., 224
flour beetle, 149–50
flower flies, 76, 239
"fly-free date," 236
food, insect contamination of, 51–52
food chain
affected by insecticides, 211–14
and phenology, 241
food choices of insects, 132
aphids, 87
apple maggot, 76–78
black swallowtail butterfly, 114
blood as food, 22, 23, 25, 92
cabbage white butterfly, 129–41
chinch bug, 220
corn rootworms, 61, 63–64
generalist feeders, 131–32, 141
grasshoppers, 153
host specific feeders, 13, 128, 138–39, 220, 266–67, 270
mosquitoes, 21, 22, 23, 25
nutrient self-selection, 148–54, 267
"malaise hypothesis," 152–53
pea aphid, 82–83
seed weevil, 180–81
Tephritidae, 70–71
toxic foods as choice, 118–19, 124, 133, 180–81
food cravings in humans, 152
Forbes, Stephen Alfred, 14–15, 65
Forbush, Edward, 156–57
Ford, Henry, 39, 162
Fraenkel, Gottfried, 133, 136, 137–38
Frazzetta, Tom, 45
Friedman, Milton, 251
Friedman, Stanley, 150
fright response, 22
fruit fly
Drosophila, 15–16, 49–60
Tephritidae (true fruit flies), 49, 69–79
Fullard, James, 148
fundatrix, 84

galls, 71, 72, 172, 173–74, 178
Galton, Francis, 53
gemmules, 53, 54
gene pool, 57
genetically modified organisms, 59, 194, 202–205
genetic drift, 57
genetics, 52–55, 56
corn rootworms, 64–65
specialization, 88
geographic isolation, 68, 70
Gerardi, Michael, 158, 161, 167
germ plasm theory, 54
gestation period, 96
Gewolb, Josh, 95
Giordano, Rosanna, 64–65
Glass, E. H., 190, 191
Glossina. *See* tsetse fly
glucosides, 135, 136
GMO. *See* genetically modified organisms
Golden Fleece Award, 257
Goodenough, Ursula, 88
gradual metamorphosis. *See* metamorphosis
grafted plants, 175, 272
Graham-Smith, George, 38

Grannett, Jeffrey, 176, 178
grape phylloxera, 170, 171–82
grasshoppers, 153
Great Wine Blight, The (Ordish), 171
greenbottle flies, 253
green stink bug, 225–26
Grimm, James, 158, 161, 167
grubs. *See* larvae
Guthion, 187
gypsy moth, 155–70
gyptol, 159–61, 166
  See also pheromones
gyroscope, 47

Haemogogus. *See* mosquitoes
halteres. *See* balancers
Harmonia axyridis. *See* Asian ladybird beetles
Harp, Joel, 153
Harpagomyia. *See* mosquitoes
Harvard University Medical School, 64
Harwood, Robert, 92
Haseman, Leonard, 102, 104, 105
hawthorn maggot, 70–71
Hazel, Wade, 120
hearing and insects, 93
Helicoverpa zea. *See* corn earworm moth
Helm, Charles, 237
hemophilia, 202
Hendrichs, Jorge, 74
heredity. *See* inheritance
herring gulls, 77
Hessian fly, 169, 233–42, 269
Hobson, Keith, 205
honeydew, 46
horses and house flies, 38–39
Horsfall, William, 25, 26–27, 32
host-marking pheromone. *See* pheromones
host specific feeders. *See* food choices of insects
hot-air balloons, 147–48
house fly, 37–48
  diseases, 38
house mosquito, 25, 29
hover flies, 93
Howarth, Francis, 251
Howlett, F. M., 262
How to Spray the Aircraft Way (Farmer's Bulletin 2062), 209
Hoy, Marjorie, 166–67
human bot fly, 254
humidity receptor, 48
Hutchinson, G. E., 139
Hybanthus shrub, 238
hybridization, 53
hybrid rootstocks of grape vines, 175, 176, 272

Icerya purchasi. *See* cottony cushion scale
Illinois Department of Agriculture, 208
Illinois Natural History Survey, 28, 199, 208–209, 210
induced secondary pests, 190–92
inheritance, 54–55, 57, 58
insect contamination of food, 51–52
insecticides, 166, 187, 192, 202, 235, 271
  banning of, 214
  contamination by, 188, 189
  dangers of, 207–18
  defense of use, 212, 214
  dieldrin, 208–209, 210, 212, 213, 214, 223–24, 271
  induced secondary pests, 190–92, 193–94, 269
  natural, 179, 194, 209
    Bacillus thuringiensis, 166, 169, 194, 201, 202, 203–205

resistance to, 33, 40–41, 45, 59, 73, 186–87, 193, 266
spraying of, 82–83, 209
timing of, 198–200
use of fake cows, 95, 272
See also DDT; noninsecticidal alternatives
insect protection against bats, 148
insects as enemies of insects, 82, 185, 247–51
instars, 116–17, 240, 244
See also molts
Instituto Columbiano Agropecuaria, 226
Integrated Pest Management, 216, 271
internal fertilization, 44
invading insects, 169–70
cabbage white butterfly, 129
Colorado potato beetle, 231
European corn borer, 195–96
grape phylloxera, 171–72
gypsy moth, 156
Hessian fly, 233–34
Mediterranean fruit fly, 72
IPM. *See* Integrated Pest Management
Ivie, G. Wayne, 181

James, Maurice, 92
Janz, Niklas, 131
Japanese beetle, 169, 207–18
milky disease, 216–17
Jeffords, Michael, 127
Jones, E. T., 234
Journal of Economic Entomology, 209
J. R. Geigy and Company, 39

Kaufmann, Tohko, 100, 153
Kearns, Clyde, 40
Kettlewell, H. B. D., 121, 122
Kitron, Uriel, 28
Klamath weed, 231–32
Knab, Frederick, 23–24
Knipling, Edward, 257, 258, 260
Koebele, Albert, 246–47
Kogan, Marcos, 237
Kosztarab, Michael, 245
Krasfur, E. S., 252
Krizek, George, 114, 117
Krysan, James, 63, 66
k-strategy. *See* parental care

Laake, E. W., 43
labella, 46
lac, 245
lady-bird beetle. *See* Australian ladybird beetle
ladybird family, 226
Lagoy, Peter, 103
L-arginine, 180
larvae, 19, 116–20, 133–34, 148–51, 268
caterpillars
black swallowtail butterfly, 115
cabbage white butterfly, 129–30, 135–36
codling moth, 184–85
corn earworm moth, 143–44
European corn borer, 196–97
evergreen bagworm, 99, 100–102
gypsy moth, 159, 162, 164
promethea, 103–104
corn rootworm grubs, 62
Drosophila, 51
maggots, 253–54
apple maggot, 73–74
Blepharoneura, 71
Euleia, 72

Euphranta toxoneura, 72
Hessian fly, 234–35
house fly, 42
Plioriocepta, 71
screwworm fly, 255
Tephritidae, 69–70
Termitorioxa termitoxena, 72
mosquitoes, 19–21
protection from predators, 116–20, 140–41, 280n2
sawfly, 72
seed weevil, 180–81
tsetse fly, 91, 95–96
See also nymphs
Lazell, James, 230
L-Canavanine, 180–81
lead arsenate, 187, 188–89, 190, 191
leaf beetle, 61, 231–32
Lenoble, Beth, 97
Leonhardt, B. A., 105
Lepidotera, 114
Levine, Eli, 66, 67
Levuana moth, 248
Liebhold, Andrew, 158
life cycles, 19
aphids, 84–85
apple maggot, 73–74
bean leaf beetles, 228
black swallowtail butterfly, 114–15
cabbage white butterfly, 129–31
chinch bug, 220–22
codling moth, 183–85
corn earworm moth, 143–45
corn rootworms, 62
cottony cushion scale, 244–45
Drosophila, 50–51
European corn borer, 196–98
European red mite, 192
evergreen bagworm, 100–102
fall cankerworm, 241–42
grape phylloxera, 173–74
gypsy moth, 159–60
Hessian fly, 234–35
house fly, 42–45
Japanese beetle, 208
screwworm fly, 255–56
sheep ked ("tick"), 97
Tephritidae, 70–71
tsetse fly, 95–97
life span
codling moth, 183
corn earworm moth, 145
Drosophila, 50
gypsy moth, 159
Hessian fly, 236
screwworm fly, 255–56
linkage (genetics), 58–59
Linnaeus, Carolus, 253
Lorenz, Konrad, 152
Losey, John, 202
"lover of dew." *See Drosophila*
Lu, Wenhua, 230
Luckmann, William (Bill), 193, 199–200, 201, 208–209
Lymantria dispar. *See* gypsy moth
Lyons, D. Barry, 158

maggots. *See* larvae
Maier, Chris, 76
"malaise hypothesis," 152–53
malaria, 18, 25, 30–34, 271
bird malaria, 24
malaria mosquitoes, 21, 24, 25, 30–34
See also mosquitoes
malathion, 33
Malayan spider, 208n2
male annihilation technique, 261–63
males

cottony cushion scale, 244
Drosophila, 50
European corn borer, 196
evergreen bagworm, 102, 104
grape phylloxera, 174
gypsy moth, 159–61
house fly, 44
mosquitoes, 23–24
promethea, 124–27
screwworm fly, 258
Tephritidae, 74–76
tsetse fly, 95
See also male annihilation technique; sterile male technique
Mallota posticata. *See* flower flies
Mangelsdorf, Paul, 63
mango shoot moths, 249
Manson, Patrick, 26
Mansonia perturbans. *See* mosquitoes
Marston, Norman, 228
Masculina arion, 251
Massachusetts Agricultural College, 156
Massachusetts Agricultural Experiment Station, 195
mating and mate selection, 269
black swallowtail butterfly, 116
Drosophila, 50–51
effected by *Wolbachia*, 64–65
evergreen bagworm, 105
fruit fly, 74–76
house fly, 44
mosquitoes, 24
screwworm fly, 258
tsetse fly, 95
use of senses, 93
See also reproductive strategies
Max 454. *See Bacillus thuringiensis*
Mayetiola destructor. *See* Hessian fly
mayolenes, 140–41
McCracken, Gary, 147–48
McLeod, D. G. R., 196
measuring worms, 241–42
Medford, Massachusetts
and European corn borer introduction, 195
and gypsy moth introduction, 156, 163
Mediterranean fruit fly, 72–73, 169
sterile male technique, 261
See also fruit fly
Meinwald, Yvonne, 120
Melophagus ovinus. *See* sheep ked ("tick")
memories of birds. *See* birds, avoidance of toxic insects
Mendel, Gregor, 52, 54, 56, 57–58
merozoites, 30–31
metabolism and malaise, 153
metamorphosis, 19
See also life cycles
Metcalf, Robert A., 44, 97, 145, 185, 225, 271
Metcalf, Robert L., 33, 44, 97, 145, 185, 225, 271
Methyl eugenol, 262–63
Mexican bean beetle, 226–27
Mexican free-tailed bats, 147–48
Meyer, Ronald, 192
microscopes and genetics, 53–54
microtome, 53–54
migration
of the house fly, 43
of monarch butterfly, 98
and natural selection, 57
migratory birds. *See* birds
milk glands
of sheep ked ("tick"), 97
of tsetse fly, 91, 95

milky disease, 216–17, 272
mimicry to avoid predators. *See* defenses of insects
Missouri Agricultural Experiment Station, 102
mites, 192
Modern synthesis. *See* neo-Darwinian synthesis
molts, 42, 95, 102, 116–17, 159, 245
monarch butterfly, 114, 131
  choice of toxic foods, 118–19, 124, 133
  impact of GMO on, 202–205
monophagous feeders, 144
  See also food choices of insects
Montana State Board of Entomology, 43
Moorfield, Herbert, 40
Moran, Nancy, 84
Morden, Robert, 100, 104, 107, 109
Morgan, Thomas Hunt, 48, 55, 56–59
Mosquito: A Natural History of Our Most Persistent and Deadly Foe (Speilman and D'Antonio), 18
mosquitoes, 17–36, 169–70
  and diseases, 25–35
  and *Bacillus thuringiensis israelensis*, 201
  sterile male technique, 261
  and *Wolbachia*, 64
moths, 168, 181
  Artona, 248
  Cactoblastis, 250
  cecropia, 108–109, 110, 168
  cinnabar, 119
  and cocoons, 114
  codling, 183–94, 261
  corn earworm, 143–54, 267
  diamond-back, 136–38
  European corn borer, 169, 195–205
  gypsy, 155–70
  Levuana, 248
  mango shoot, 249
  peppered, 120–22, 266
  promethea, 103–104, 122–23, 124–27, 168
  silkworm, 107–108, 131
  tobacco budworm, 147
  tobacco hornworm, 13, 85, 133, 139
mouthparts, 46, 115
Müller, Fritz, 118
Muller, Herman, 56
Müller, Paul, 39
Müllerian mimicry. *See* defenses of insects
Musca domestica. *See* house fly
mustard family of plants, 129–30, 135, 136–38
mutations
  black peppered moth, 121–22
  Drosophila, 57–58
  plant and insect interaction, 134–35, 139
myrosin, 136, 137–38
Myxomatosis, 34–35

Naflus, Donald, 249
nagana, 92, 96
naled, 263
Nanney, David, 53, 55
National Geographic (magazine), 147
National Research Council Corn/Soybeans Study Team, 225
natural insecticides. *See* insecticides
natural selection, 35–36, 41–42, 56, 178
  and corn rootworms, 66, 67–68
  and European grapes, 174

examples of, 265
genetics, 57
mating strategy, 74–76
and parthenogenesis, 88
in plants, 134–35
Nature Wars (Winston), 41, 185
nematode worms, 26, 35
neo-Darwinian synthesis, 56, 57
"New Jersey state bird," 18
Newnham, A., 117
New World screwworm. *See* screwworm fly
New York State Agricultural Experiment Station, 190
nicotine, 179, 209
nightshade family of plants, 133
nocturnal insects, 123, 145, 148, 268
noninsecticidal alternatives, 73, 271
fungus, 83
See also insecticides
northern corn rootworm, 62, 63, 64–65, 66
See also corn rootworms
Novak, Robert, 28–29
noxiousness and avoidance by predators. *See* defenses of insects
nutrient self-selection. *See* food choices of insects
nutritional wisdom, 148–54
nymphs, 19, 133, 192
chinch bug, 221–22
cottony cushion scale, 245
grape phylloxera, 173, 176, 178
See also larvae

obligatory diapause, 106–107
See also diapause
odors. *See* smell and insects
Ohio Agricultural Experiment Station, 190
Oldroyd, Harold, 254
Old World screwworm, 254
oligophagous feeders, 144
See also food choices of insects
olive fruit fly, 78–79
Opler, Paul, 114, 117
Ordish, George, 171
"organic" farming, 188
oriental fruit fly, 74, 261–64
Origin of Species, The (Darwin), 48
osmeterium, 119–20
Ostrinia nubilalis. *See* European corn borer
overwintering, 29
See also diapause
ovipositor, 70, 77, 78
Oxford University, 117, 121

Packard, C. M., 223
Pal, R., 40
pangenesis hypothesis, 52–53
Papaj, Daniel, 76
Papilio polyxenes. *See* black swallowtail butterfly
parasites, 166, 185, 242, 248, 249, 268, 269
that rescue plants, 146
See also biological controls
parathion, 190
parental care, 79, 85, 89, 269–70
k-strategy, 85, 96
r-strategy, 85
and tsetse fly, 91–98
Parker, R. R., 43
parsley caterpillar. *See* black swallowtail butterfly
parsley plants, 181
parsleyworm, 181
parsnip webworm, 181

parthenogenesis, 84, 85–89, 173–74, 242
parturition, 95
pathogens, 166
pea aphid, 82–83, 86
peach fruit fly, 262
pea plants, 54, 57–58, 180–81
Pedigo, Larry, 14
penis, 44
peppered moth, 120–22, 266
pesticides. *See* insecticides
pest insects, definition, 14–15
pest management, 192
Petty, Howard, 201
phenology, 236–39, 267
  and plants, 238
pheromones
  host-marking, 78–79
  sex, 50, 62, 84–85, 105, 145, 183, 196, 244
    gyptol, 159–61
    male released, 74
    methyl eugenol, 263
    and virgin promethea, 125
  synthetic gyptol, 161, 166
  trail, 161
  warning, 161
photoactive toxins, 181
physiological defenses against predators, 181
Phytalmia, 71–72
Pickett, L. J., 84
Pieris rapae. *See* cabbage white butterfly
Pimentel, David, 169
pipevine swallowtail butterfly, 122–23, 125, 127
pitcher plant mosquito, 25
placenta, 91
Planchon, Jules, 175–76
plankton as food, 21, 25
plant and insect interaction
  biochemical defense against insects, 134–41, 146–47, 178–82, 240
  host plant selection, 138–39
  insect circumvention of defenses, 139–40, 180–81, 182, 249–50, 267
plant lice. *See* aphids
plant parasites, 244
plant preference for food. *See* food choices of insects
plants and phenology, 238
Plasmodium, 30–31, 33–34
  See also malaria
Plioriocepta, 71
polyphagous (generalist) feeders, 131–32, 144, 196, 225
  See also food choices of insects
pomace flies. *See Drosophila*
Pomonella. *See* apple maggot
Popillia japonica. *See* Japanese beetle
Population, Resources, Environment (Ehrlich and Ehrlich), 214
potato, 229–31
Potter, Daniel, 101
Poulton, E. B., 117
predators, avoiding. *See* defenses of insects
predators of insects, 147–48, 166, 185, 242, 269
  bats, 147–48, 186, 268
  See also biological controls
Price, Peter, 232
primary plant substances, 179
Proceedings of the National Academy of Sciences (2002), 140
Prokopy, Ronald, 74, 76–77, 78
promethea, 103–104, 122–23, 124–27, 168

propoxur, 33
Protocalliphora, 254
Proxmire, William, 257
psoralens, 181–82
Psorophora horrida. *See* mosquitoes
Ptychomyia, 248
pubescence, 105
pupae, 19
  apple maggot, 74
  black swallowtail butterfly, 114–15, 120
  cabbage white butterfly, 130–31
  choice of locations, 102–104
  cocoons and chrysalis, 102–104, 114, 130
  codling moth, 185
  corn earworm moth, 144–45
  corn rootworms, 62
  Drosophila, 51
  European corn borer, 196
  evergreen bagworm, 102
  gypsy moth, 159
  Hessian fly, 235
  house fly, 42–43
  mosquitoes, 21–22
  promethea, 103–104
  protection from predators, 120, 268
  screwworm fly, 256
  tsetse fly, 96
puparium, 43
pyrethrum, 179, 209

Quaintance, A. L., 69

rabbits, diseases of, 34–35
"Raison d'être of Secondary Plant Substances" (Fraenkel), 138
Ratzka, Andreas, 137–38
Raupenleim, 164
Raven, Peter, 139
recessive allele, 54
red-banded leaf roller, 190, 191
red maple and fall cankerworm, 241–42
red-spotted purple butterfly, 122
Reichert, Susan, 153
reproductive cycles. *See* life cycles
reproductive isolation, 68, 70
reproductive strategies
  of aphids, 84–89
  of chickadee, 85
  grape phylloxera, 173
  of tsetse fly, 95–97
  See also mating and mate selection
resistance to insecticides. *See* insecticides
retrogressive evolution, 244
Riley, Charles Valentine, 172–75, 178, 243, 246–47
"ripple effect," 251–52
root galls, 172
Rosenthal, Gerald, 180
rosy apple aphid, 84–85, 86
rotenone, 179
Roth, Louis, 270
Rothschild, Miriam, 240
Roundup, 202
roundworms. *See* nematode worms
Rousseau, Henri, 176
r-strategy. *See* parental care

Sabrosky, Curtis, 167
Sacred Depths of Nature, The (Goodenough), 88
St. John's wort. *See* Klamath weed
saltationists, 56
salt marsh mosquitoes, 17–18
Sanderson, Dwight, 207

Saunders, Aretas, 37
sawfly, 72
scales
commercial use, 245
cottony cushion scale, 166, 193, 242, 243–52
scarab beetles, 250–51
Schaus's swallowtail butterfly, 117
Schneider, John, 241
Schoonhoven, Louis, 131
Science (journal), 95
Scientific American, 250
Scott, Thomas, 210
screwworm fly, 253–64
sterile male technique, 258–61
seasonal patterns of development, 236–39
secondary pests caused by insecticides, 190–92, 193–94, 271
secondary plant substances, 134–35, 139, 179
seed weevil, 180–81
selectionists, 56
senses of insects, 93
sex-linked inheritance, 58
sex pheromone. *See* pheromones
sexual differentiation of evergreen bagworm, 102
sheep ked ("tick"), 97
Sheldon, Joe, 239
shellac, 245
sickle-cell anemia, 34
Siegler, E. H., 69
Silent Spring (Carson), 205, 211, 214, 215, 272
silk
butterflies, 115
evergreen bagworm, 100–101, 102
gypsy moth, 155–56, 161–62
silk, corn, 144
silk moths and gypsy moth, 251
silkworm, 107–108, 131
sinalbin, 136, 137
sinigrin, 136, 137–38
size as a factor in choice, 77
sleeping sickness, 92, 93, 96
Smedley, Scott, 140
smell and insects, 93, 119–20, 134, 179, 267
See also pheromones
Smith, Edward H., 172
Snodgrass, Robert, 83, 92
soybeans, 65–66, 67, 198, 224–25
ecosystem, 228–29
and GMOs, 202
insect predators, 225–28, 237–38
specialization and parthenogenesis, 88
specialized food choices. *See* food choices of insects
species
creation of new, 68, 70, 266
variety of, 35
Spencer, Warren, 51, 55
sperm, 44–45, 50, 76, 95, 258
fall cankerworm, 242
spice bush swallowtail butterfly, 122, 123
Spielman, Andrew, 18, 27, 29, 32
sporozoite, 30–31
spraying insecticides. *See* insecticides
squash bugs, 131
Steiner, L. F., 261
stem mother, 84, 86, 173, 174
sterile male technique, 258–61, 272
Sternburg, James, 40, 103, 109, 116, 125, 126, 127, 168
Stiles, C. F., 222
stomach poisons, 191
stowaway insects. *See* invading insects

Strickberger, Monroe, 53
summer diapause, 107
See also diapause
supernormal stimulus, 77
survival of the fittest. *See* natural selection
swarms of mosquitoes, 23–24
synchronization of hatching, 110
synchrony, 236–39, 267, 268
synthetic insecticides. *See* insecticides

Tachinidae family, 167
tachinid fly, 248, 249
Tanglefoot, 78, 164
tassel ratio, 199–201
taste and insects, 46, 77, 139, 267
telescoping of generations, 87
temperature. *See* weather
Tennessee Valley Authority, 31
teosinte, 63
Tephritidae. *See* fruit fly
termination of diapause, 108–11
See also diapause
terpenoids, 146
tetracycline, 65
Thacker, Robert, 152
Thompson, J. N., Jr., 51
Thorsteinson, Asgeir, 136–38
Thyridopteryx ephemeraeformis. *See* evergreen bagworm
tiger swallowtail butterfly, 117, 123
Tilman, David, 240
timing in relation to insects, 107–108, 236–39, 272
tobacco budworm moths, 147
tobacco hornworm moths, 13, 85, 133, 139
tomato fruitworm. *See* corn earworm moth
Tothill, J. D., 248
toxic foods as choice. *See* food choices of insects
toxicity of insecticides, 210
toxins, 137–38, 181, 198–200, 255
protein, 166, 169, 201
trail pheromones. *See* pheromones
Trichogrammatidae, 185
Trouvelot, Leopold, 156
trypanosomes, 92–93
tsetse fly, 91–98, 269
tumblers. *See* pupae, mosquitoes
tumorlike swellings. *See* galls
two-spotted spider mite, 192

"undertaker" insects, 253–54
University of California, 176, 243
University of Florida, 166
University of Guam, 249
University of Illinois Agricultural Experiment Station, 224
University of Illinois Department of Crop Sciences, 145
University of Illinois Department of Entomology, 14, 40, 71, 203
University of Illinois Department of Plant Biology, 238
University of London, 136
University of Massachusetts, 156
University of Minnesota, 198
Uranotaenia sapphirina. *See* mosquitoes
US Agricultural Research Service, Entomology Research Branch, 257
US Department of Agriculture, 144, 212, 222, 261
Bureau of Entomology and Plant Quarantine, 83, 156, 256
Farmer's Bulletin, 69, 209, 217, 223
Plant Pest Control Division, Agricultural Research Service, 208

US Environmental Protection Agency, 214
US Food and Drug Administration, 52, 189
uterus, 91, 97

van den Bosch, Robert, 216
Vanderzant, Erma, 150
Variation of Animals and Plants under Domestication (Darwin), 52
vectors. *See* diseases
vedalia beetle, 193, 247–48
Vedalia cardinalis. *See* vedalia beetle
Verschaffelt, E., 135–36
vetchworm. *See* corn earworm moth
viceroy butterfly, 114, 124
Virginia Polytechnic Institution, 245
virus. *See* diseases
vision and insects, 93, 95
Vitis vinifera, 171
von Tschermak-Seysenegg, Erich, 54

warning pheromones. *See* pheromones
wasps, 185, 249
Wassenaar, Leonard, 205
Waterhouse, D. F., 250–51
weather, 100, 107, 108–10, 132
Webster, F. M., 63
weed, definition, 15
Weinzierl, Rick, 145
Weis, Arthur, 139
Weismann, August, 54
West, David, 120
West, Luther, 38
Westbrook, John, 147–48
western corn rootworm, 62, 63, 64–65, 67, 170
  See also corn rootworms
West Nile Fever, 18, 27–29
wheat
  replaced by soybeans, 231
  timing of sowing, 236
  winter wheat, 234
white-eyed male *Drosophila*, 57–58
WHO. *See* World Health Organization
Wigglesworth, Vincent, 47
Willis, Edwin, 270
Wilson, Edward O., 270
Windecker, W., 119
wind speed indicator, 48
wine and grape phylloxera, 171–82
wings
  number of, 47
  winged and wingless generations, 84, 173–74, 242
  See also flight of house fly
Winston, Mark, 41, 185, 187
winter "sleep." *See* diapause
winter wheat, 234
Wolbachia, 64–65
Wolcott, G. N., 270
World Food Prize, 260
World Health Organization, 32, 33
Worthley, L. H., 196
wrigglers. *See* larvae, mosquitoes

xanthotoxin, 181

Yack, Jayne, 148
Yamamoto, Robert, 133
yellow fever, 18, 25–27
yellow-fever mosquitoes, 24

Zangerl, Arthur, 203–204
Ždárek, Jan, 91, 95
Zeidler, Othmar, 39
Zuckerman, Larry, 229